Investigating How and Why Motor Vehicle Crashes Occur

Decades of research reveal that human error is a significant contributor to vehicle crashes, accounting for over 90% of incidents. Humans possess inherent physical, visual, and cognitive limitations that can lead to accidents on the road. This book draws on the author's 35 years of experience investigating approximately 500 motor vehicle crashes to explore these limitations and their role in causing accidents.

In this title, ten specific crashes, each illustrating how human limitations contribute to accidents, are presented and analyzed. Technical solutions such as automated emergency braking are offered for each crash type. From passenger car drivers to pedestrians, motorcyclists to truck drivers, each chapter investigates the unique circumstances and human factors at play. In each case, the interdisciplinary science of human factors that applies is explained, covering issues such as visibility with low beam headlights, glare, fatigue, information processing demands, and false expectations. Including a final chapter on the author's experiences acting as a human factors expert witness in court, this book delivers to the reader a vital overview of human error in vehicle accidents.

Investigating How and Why Motor Vehicle Crashes Occur: To Err Is Human is an essential read for practitioners of human factors and ergonomics, occupational health and safety, automotive engineering, civil engineering, road and transport design, collision investigation and the law.

Investigating How and Why Motor Vehicle Crashes Occur

To Err Is Human

Alison Smiley

CRC Press
Taylor & Francis Group
Boca Raton London New York

CRC Press is an imprint of the
Taylor & Francis Group, an **informa** business

Designed cover image: iStock by Getty Images

First edition published 2026
by CRC Press
2385 NW Executive Center Drive, Suite 320, Boca Raton FL 33431

and by CRC Press
4 Park Square, Milton Park, Abingdon, Oxon, OX14 4RN

CRC Press is an imprint of Taylor & Francis Group, LLC

ISBN: 978-1-032-93186-9 (hbk)
ISBN: 978-1-032-91247-9 (pbk)
ISBN: 978-1-003-56480-5 (ebk)

DOI: 10.1201/9781003564805

Typeset in Times
by Deanta Global Publishing Services, Chennai, India

Dedicated to

Thomas Smahel, from assistant to partner, for 26 years of enjoyable and productive collaboration.

Contents

About the Author

Dr Alison Smiley served as president (1984–2021) and principal consultant of Human Factors North (HFN) Inc. in Toronto, Canada, holding a PhD (1978) in Systems Design Engineering with a Human Factors specialization. As the founding Chair of the Canadian College for the Certification of Professional Ergonomists (CCCPE), she brings 40 years of expertise in measuring driver performance affected by sign design, distraction, fatigue, drugs, and headlight design. Dr Smiley has taught human factors courses to traffic engineers, highway designers, university students, and accident reconstructionists throughout Canada and the United States, while contributing to various manuals and books, including the Ontario Traffic Manual and the Geometric Design Guide for Canadian Roads. Her extensive experience includes conducting investigations and writing reports for over 500 legal cases involving motor vehicle crashes, and she has provided expert testimony in court more than 20 times across both Canada and the United States.

Preface 1

Road safety became a salient public issue in North America 60 years ago with the damning of the automotive industry by Ralph Nader, author of *Unsafe at Any Speed*. Studies in response to this confirmed that the human factor is the leading cause of crashes and is exclusively or partially responsible for over 90% of them.

It took the safety community close to half a century to realize, as Dr Smiley states in the subtitle of her book, that "To err is human". Human failures such as distraction, fatigue, and human limits, such as in visibility and perception-reaction time, can best be overcome not by changing human behaviour, but by changing the circumstances in which behaviours and related accidents occur. As she notes in her introduction, changing human behaviour is not rocket science – it is a **lot** harder. And she should know. This book is built on Dr Smiley's background in road safety, which offers a unique combination of scholarly research publications, practical contributions, and extensive expert witness experience in road safety. This book provides real traffic crash scenarios, their relevant engineering, psychological and scientific context, and the benefit of theoretical and practical expertise in generalizing from these individual cases to wide-ranging applications to improved road safety.

This book is not another scientific, professional, or educational textbook. It is different from any other book I know on the subject. It is on the one hand a very personal and highly readable collection of professional experiences in investigating and analyzing a set of ten road crashes, and on the other hand a very professional exposition in a layperson's language of the science involved in their occurrence.

Dr Smiley is a familiar and highly respected name in the scientific community that focuses on road safety. In her exposition of the ten cases, she illustrates the roles that visibility and illumination, and lesser known but critical visual issues such as glare, contrast, and contrast sensitivity under various conditions, and the role that "expectancy" plays in either enhancing or reducing visibility.

While reading this book, I kept asking myself who is the intended audience for such a book. My conclusion surprised me: almost anyone. From the scientifically curious academics to the professionals working in the field, to the people who wonder why and how accidents they either experienced or heard or read about happened. And that, I think, covers almost all of us.

Dr David Shinar
Road Safety Expert

Preface 2

Dr Alison Smiley, one of Canada's leading human factors specialists, offers us an overview of the human factors that lead to motor vehicle accidents. She is motivated by a deep curiousity for understanding the human limits that lead to errors, injuries, and deaths – and to reduce their devastating impacts. It is also inspired by her desire to describe the fascinating interdisciplinary science of human factors.

Internationally recognized, Dr Smiley has over 40 years' experience in the field. An adjunct professor at the University of Toronto, she has taught university students, traffic engineers, highway designers, and police accident reconstructionists around the world. She is the editor of and author of five chapters of *Human Factors in Traffic Safety* (Lawyers & Judges Publishing) for human factors engineers and other related professionals. Her work on human factors is adopted in publications relied on by government bodies in Canada and the United States.

Dr Smiley has been accepted by courts in Canada and the United States as an expert witness. She has been qualified to testify on human factors issues in over 20 serious accident cases, and in three inquiries. She has investigated and articulated her conclusions on the issues in over 500 cases.

As a personal injury trial lawyer, she was my first choice whenever I needed an expert human factors engineer to understand and advance my client's case. And, when my opposing counsel had retained Dr Smiley before I could, I knew I had a challenging case in front of me!

This insightful text contains a wealth of information presented through a fascinating review of her investigations, reporting, and testimonies, for ten of her most interesting and illustrative cases. This is the first book I know of devoted to an in-depth analysis of individual motor vehicle accidents, focusing on the contributing human factors, including information overload, expectations, and perception-reaction time, among other aspects.

Investigating How and Why Motor Vehicle Crashes Occur: To Err is Human is intended for a broad audience, including those who are curious about the mystery of their own collisions, as well as those with professional expertise in the field. Dr Smiley's book will also be an essential educational tool for accident investigators, insurance adjusters, police, lawyers, technical students in related fields, engineers, and psychological forensic specialists.

No one with an interest in this complex and constantly changing field can afford to be without this book.

Roger G. Oatley
Personal Injury Lawyer

Acknowledgements

It takes a village to write a book! There are many people to thank. These include my esteemed colleagues at HFN – especially Thomas Smahel, who read and commented on key portions of the book and contributed most of the photographs and collision diagrams. Dr Rudin-Brown read and commented on numerous chapters. HFN partners generously provided funds for assistants Olivia Dunford and Alia Mohamed Ahmed Galal.

Several lawyers provided helpful advice and no doubt assisted in avoiding expensive lawsuits! Bob Munroe, Roger Oatley, Alan Rachlin, James Mays, Alan Bromiley, Lawrence McRae, Philip Buckingham, Dianna Hwang and the Honourable Sheila O'Connell were most encouraging and helpful in reviewing my work from a legal perspective.

There were a number of others representing both experts in related fields and the broad general audience that I hope to reach, who read and commented on one or more chapters. These included: Geni Bahar, Peter Rudin-Brown, Ivan Jaye, Ingvar Thor Jonsson, Tom Robinson, Lennox Shantz -Wagman, Peter O'Brien, Dr. Brent Kilborn, Almis Ledas, Dr. Tatiana Freire-Lizama and Dr. Don Donderi (former HFN partner).

I also thank Jason Zsoldos, who volunteered to be identified and to share the actual details of his tragic crash, facilitated by his lawyer James Mays. Jason's story is told in Chapter 4.

Dr Maurice Masliah very kindly made a graduate course project out of the chapters, and by that means I was able to get insightful feedback from students in his graduate class MIE1414 at the University of Toronto.

I thank James Hobbs and Becky Lane of CRC Press, and Melwin Ignatius of Deanta who made the process of copyediting, revising and publishing a book relatively painless. In particular, I thank James Hobbs whose first very enthusiastic response to my proposed book was enormously motivating and Melwin Ignatius for his patience.

Last but not least, I thank my family for their never-ending support, especially my husband, Bob Shantz, who encouraged me in numerous ways, read the entire book at least twice, made many useful comments, and put up with my talking about it for the 2 years it took to write it, my daughter Sarah Smiley who came from Iceland in part to finish the last bits and pieces, and my daughter-in-law, Cindy Wagman, who (together with my friend, Susan Rynasko,) for the job of selecting the book cover image.

Finally, I take responsibility for all errors and omissions, and I am sure there are some, for to err is human!

1

Introduction

Introduction

To err is human. In other words, people make mistakes. As an ergonomist, my work involves understanding human limits that lead to errors, injuries, and deaths. I am particularly interested in errors made by road users. The aim of my profession is to reduce errors and lessen the impact of errors when they do occur. This work led me to become a human factors expert witness in legal cases involving motor vehicle crashes in Canada and the United States. In this book, I share some of the remarkable stories about crashes I have investigated.

I also describe some of the fascinating interdisciplinary science called "human factors" that explains why road users make the mistakes they do. The question I attempt to answer in legal cases is: what behaviour led to this crash? In many cases we have a seemingly ordinary driver, driving along, blissfully unaware of the fate that awaits. Within seconds, his or her life is turned upside down in a motor vehicle crash.

What egregious error could they possibly have made to deserve the horrible situation in which they and the others involved find themselves? Humans have built-in physical, visual, and mental (cognitive) limitations that can get them into trouble on the road, and human error contributes to about 90% of crashes. That does not mean, as many think, that the best way to deal with human limitations is through education or enforcement. It has been said, by those who know, that changing human behaviour is not rocket science. It's a lot harder! Engineering changes (the addition of shoulder edge rumble strips on highways, or replacement of a standard four-leg intersection with a roundabout, for example) are often more successful.

Intended Audience

This book is intended for a broad audience, including anyone intrigued by motor vehicle crashes and how they happen. This audience includes people

DOI: 10.1201/9781003564805-1

who have been involved in crashes themselves, police who make the initial report, lawyers who sue and defend crash participants, investigators who gather witness evidence, and insurers who cover costs. It also includes the wide variety of expert witnesses who participate in crash investigations, including reconstruction engineers, who deal with such issues as speed at impact and crash dynamics; mechanical engineers, who report on the condition of the brakes and tires; forensic experts, who determine alcohol and drug levels; and psychologists and physicians. These professionals provide expert opinions themselves and will be interested in the basis of the opinions from other relevant experts. To encompass this broad audience, I write in as accessible a way as possible.

For many years I taught courses in human factors. One of the projects involved students in investigations of past crashes. They then participated as human factors experts in a moot court. I was the judge. Role play is not usual in engineering courses, and I was not sure how the students would react to being asked to do so. However, when I saw the "lawyers" arriving in suits, I was sure the moot court was going to be a success! It was always one of the most popular projects in the course. Based on that experience, I believe that teachers and senior students at every level and in many areas of study will find this book both intriguing and beneficial.

My hope is that this book will enable some understanding of human limitations, and that some road users will therefore avoid making mistakes that could lead to serious injury or death. People involved in crashes, particularly those in which their loved ones are injured or killed, and especially those for which they were at fault for the crash, can suffer from guilt or trauma for many years. I hope that some of these people might be able to forgive themselves after realizing that their error was a result of normal human limitations.

What Is Human Factors About?

When I first started acting as an expert witness in court, most lawyers were unfamiliar with human factors/ergonomics as an area of expertise. Many lawyers still are. Prior to the use of human factors experts, any human factors done in the analysis of the cause of a crash tended to be at a very simple level. Such analysis mostly involved an estimation of reaction time to a hazard on the roadway. Most of the cases I did in the 1980s and 1990s did not include a human factors expert on the other side. This gave my side an advantage and made my job less difficult. Now it is much more likely that the other side also has an expert witness in human factors, and so giving evidence is more challenging. Even so, despite the legal community being more informed about the value and relevance of human factors, almost every legal report I write includes a paragraph defining human factors, even if not asked to do so. Here's an example:

> "Human factors" (also known as ergonomics) is the scientific discipline concerned with the understanding of interactions among humans and other elements of a system. The aim is to improve human well-being and overall system performance. Human factors draws on knowledge from the human sciences, including physiology, psychology, and kinesiology (study of movement) and applies it to engineering design. The aim is to reduce errors and injury.

Human factors can be applied to every tool, piece of equipment, and workplace. It can be applied to the design of a toothbrush, to the design of roadways and traffic signs, and to the design of an international space station. In this book, human factors is applied to a wide variety of road users, vehicles, and driving and walking environments.

Human factors practitioners tend to specialize in one of two areas. Some focus on human physical limitations, generally called ergonomists. These individuals deal with the "neck down" and typically have a background in kinesiology and physiology. Then there are those who focus on the "neck up", that is, perceptual and cognitive areas. This latter group are generally called human factors specialists. I am one of these. Traffic crashes typically involve perceptual issues, such as limitations in seeing distance at night, the impact of glare, sensitivity to contrast, hearing threshold, and cognitive issues such as speed of information processing, short-term memory, effects of alcohol and drugs, fatigue, distraction, and expectation on perception-reaction time (PRT). Both "neck up" and "neck down" practitioners can be certified as "ergonomists".

What Is This Book About?

During my career, I have investigated over 500 road crashes across Canada and the United States. For this book, I have selected cases representing a broad range of crash types. The 10 chapters that follow each deal with an individual case. One involved a young man driving two friends home late one rainy night on an unfamiliar road. Confused by the signing and the road layout, the driver realized too late that the road he was following did not go straight but curved to the right. The result of his delayed reaction was that he was going too fast to make the curve. He ran off the road into a large rock resulting in the deaths of the driver and his two passengers.

A second involved a nurse on her way to work who entered the intersection just as she had a green light. She struck a teenager crossing at a red light.

The nurse's view towards the pedestrian was blocked by an SUV in the left turn lane.

I deal with crashes that are related to fatigue, visibility, visual search, alcohol, drugs, expectancy, and road design. Typically, each case involves a number of human factors issues. I have selected several factors in each case and explored the underlying science in detail. Each case is described using the same format:

Introduction

Overview of the type of crash and the outcome.

The Collision

A description of the crash, for example:

> It was a dark, rainy night in a rural area. Ms Jones was driving at the speed limit of 80 km/h when she suddenly encountered a pedestrian in dark clothing crossing the highway. She swerved onto the shoulder but was unable to avoid striking the pedestrian, resulting in his death.

The Mandate

This lays out the purpose of the human factors expert report, for example:

- Provide an opinion on whether a reasonably cautious driver could have reacted in time to avoid the pedestrian, given the visibility at the time.

The Investigation

This usually involves a re-enactment of the collision to look at various issues, including:

- seeing distance to a hazard
- how likely a driver is to notice the hazard
- whether the presence of streetlighting would have made the crash less likely

The aim of the re-enactment is to allow the investigator to see the scene as the driver, pedestrian, or bicyclist involved did. Compromises must be made with respect to the reality, and of the accuracy of the re-enactment, to be sure that one does not repeat the crash! For example, unless there is no traffic, the assistant playing the role of the pedestrian stands on the edge of the road rather than in the centre of the road while measurements are taken of seeing distance.

The Science

This articulates some of the human factors issues, usually numerous, that would have been in play. The scientific literature is consulted. For example, there are studies of how far away one can detect a pedestrian in a white vest using low beam headlights on country roads.

Conclusions

These are based on the evidence, investigation, and analysis.

Countermeasures

This discusses safety measures which target the crash types.

In the last 40 years, a great deal of research has gone into more sophisticated statistical analysis of how highway design changes affect safety. An outstanding example is the replacement of standard intersections with modern roundabouts. In the right circumstances, these can reduce injury crashes by 70% (Persaud et al., 2000).

Another trend is the development of driver assistance systems, such as Forward Collision Warning Systems, which can sense slow or stopped traffic in the path ahead. Some advanced systems using infrared light can detect pedestrians, cyclists, or large animals and warn the driver (Government of Canada, n.d.). Besides warning, there are active systems that apply the brakes or steer back into the lane when the driver drifts onto the shoulder or over the centreline.

Such countermeasures are highlighted in each chapter.

References

This book is not intended to be an academic treatise and therefore is not fully referenced. The more important references to published books and papers are provided.

Maintaining Privacy

In order to tell some of these stories and maintain privacy, the details are changed, and in one case, the facts of one case are mixed in with those of another. It can be said that, like some movies, these stories are based on real-life events. It all happened, but perhaps not with quite the same combination of facts or timeline. Identifying names, dates, and locations have been removed unless otherwise agreed.

Acting as an Expert Witness

The last chapter describes my experience of acting as an expert witness. This includes how lawyers identify me and other potential experts, the courtroom qualification process as an expert, and the three parts of giving evidence: the examination-in-chief, the cross-examination, and the re-direct. About 95% of cases settle out of court, and much drama can be present in the remaining 5%. The last chapter describes some of the experiences I have had, both pleasant and unpleasant, in some of the 25 cases in which I have been involved that have gone to trial.

Accident vs Crash or Collision

In writing this book, I struggled over the use of the word "accident" versus "crash" or "collision". The traffic safety community wants to call such an event a "collision" or a "crash", arguing that the word "accident" implies that nothing can be done about it. Given that there are many effective programmes for educating drivers, many effective enforcement programmes, and many effective solutions or treatments for road design and traffic operations, it is manifestly true that there is much that can be done to prevent crashes.

On the other hand, I and many others feel that a crash or collision should be called an accident, in that no one intends to have one. Also, I feel "accident" is a less technical term, perhaps more appropriate for a book aimed at a broad audience including both professionals and non-professionals. Out of respect for my colleagues in traffic safety, I use the terms "crash" or "collision".

References

Government of Canada. (n.d.). What you need to know about driver assistance technologies. https://tc.canada.ca/en/road-transportation/driver-assistance-technologies/what-you-need-know-about-driver-assistance-technologies#_Collision_warnings (last accessed April, 2025).

Persaud, B. N., Retting, R. A., Garder, P. E., & Lord, D. (2000). Crash reductions following installation of roundabouts in the United States. *Insurance Institute for Highway Safety, 15*, 540–542.

2

Pedestrian Lying on the Roadway

Introduction

This chapter describes an accident in which a driver encountered a pedestrian lying on the roadway in a poorly lit area at night. The driver was unable to stop in time and struck the pedestrian. In The Science section of this report, the effects of contrast, headlighting, streetlighting and expectancy on seeing distance are discussed.

The Collision

According to the police report, Mr James Smith (aged 57 years), driving a Toyota Corolla, struck a pedestrian on Terminal Street in an industrial area just before midnight in mid-September. The street had two lanes in one direction, and one lane in the other. Mr Smith was driving in the centre lane. The posted speed limit was 50 km/h. There was little traffic at that time of night.

The pedestrian, Mr Solomon (aged 25 years), was very depressed and had been drinking heavily. He was walking home from a bar. For reasons that are not clear, when found, the pedestrian was said to be lying parallel to the roadway in a manner that appeared almost suicidal. Suicide by motor vehicle is more common than you might think. A recent study estimates that fully 12% of motor vehicle crashes involving men, that resulted in fatalities, were determined to be suicides (Skyving et al., 2023).

Shortly after Mr Solomon lay down, along came Mr Smith who did not see him until after he struck him. And even then, Mr Smith thought he had struck a garbage bin or something similar. He said he did not realize it was a person and he did not stop. Mr Solomon was seriously injured.

There was streetlighting on one side of Terminal Street. The police reported that the streetlight nearest to Mr Solomon was burnt out. See Figure 2.1.

 DOI: 10.1201/9781003564805-2

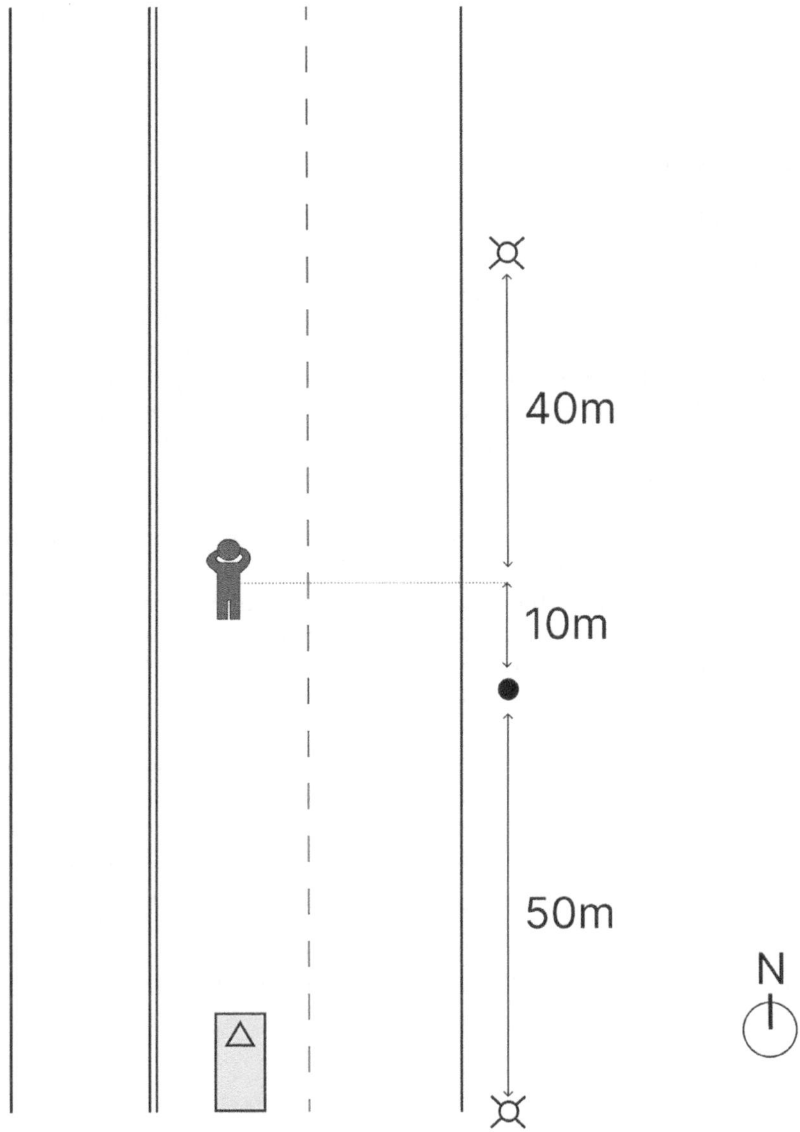

FIGURE 2.1
Pedestrian lying in road 40 m from nearest operating streetlight.

The Mandate

Mr Solomon sued Mr Smith, claiming that he should have been able to see him (Mr Solomon) in time to stop before striking him. My job was to determine whether the collision could have been avoided by a reasonably alert and cautious driver.

The Investigation

Purpose

A central question for the investigation was to determine at what distance a pedestrian in dark clothing, such as Mr Solomon was wearing, could be seen, using low beam headlights in a similarly streetlight-lit area, at night. This called for a re-enactment of the circumstances of the collision, to the degree that they were important determinants of visibility.

Most people would think, correctly, that the easiest way to do a re-enactment would be to visit the actual site. This was not possible for this collision because the nearest streetlight to the pedestrian was not operating at the time of the collision but had been repaired by the time we were able to do our site visit. Consequently, the re-enactment had to take place in similar lighting conditions to those of the collision but not at the original site, that is, in a streetlight-lit area but where the nearest streetlight was not working. It also had to take place at night at least an hour after sunset. These requirements are discussed below.

In choosing a re-enactment location, besides being able to view the pedestrian at the same distance from a working streetlight, we needed to view him or her under no traffic conditions (i.e. no glare), similar road environment lighting, and a similar row of streetlights on one side of the roadway with one not working. (In my experience, non-working streetlights are not difficult to find. And, in my experience, they are frequently a factor in pedestrian accidents.) In addition, there were factors to consider that were not affected by the location, such as headlighting. And last but not least, it was important that the re-enactment be done safely.

Time of Day

Obviously, time of day is a major factor affecting visibility distance and is determined by latitude, and day of the year. The pedestrian collision we are considering in this chapter occurred at midnight, in mid-September, at a location where sunset was around 7:15 p.m. and the end of civil twilight about 7:45 p.m. Thus, the collision occurred 4.75 hours after sunset. Sunset occurs when the upper edge of the sun touches the horizon. Outdoor activities are still possible, but colours are difficult to see. About an hour after sunset, it can be considered to be fully dark for the purposes of a site visit to measure visibility distance with headlights and streetlights. The site visit does not have to occur on the anniversary date of the collision; although this may be desired by lawyers for the purpose of credibility, it is not necessary to do for the sake of accuracy, but it does have to be done when the light level is the same as on the day of the collision. The site visit was done in early January on a day when sunset was around 5:05 p.m. The light level was approximately the same as that at the time of the collision, that is, fully dark, by 6:05 p.m.

As long as traffic volume was equivalent to that at the time of the site visit, any time after 6:05 p.m., about an hour after sunset, would have been considered fully dark and suitable for a site visit (and much more convenient than the time of the collision (midnight)). Our site visit was conducted at 9:00 p.m.

The timing of the site visit would have been more limited had the collision occurred near sunrise/sunset, when light level and therefore visibility distance change rapidly, and when one may have to consider not only the altitude of the sun but its azimuth, that is, how far south the sun has moved. One must also be careful with calculations in the case of spring and fall when the time changes.

Some people think that the phase of the moon should be considered as a factor affecting visibility distance. However, research has shown that drivers using headlights are adapted to light levels that are high enough that additional light from the moon is of no consequence (Olson et al., 1990 from Smiley et al., 2016).

Besides time of day and streetlighting, factors that were considered as being important to the visibility distance were headlighting, glare from oncoming vehicles' headlights, other lighting in the road environment (i.e. ambient lighting), the reflectivity and contrast of the pedestrian's clothing against the background in which he was seen, and position of the pedestrian.

Headlighting

When streetlighting is minimal or nonexistent drivers must rely on their headlights. The most basic question about headlights is: were Mr Smith's headlights on? Yes – the police observed this when they arrived at the scene. Were the headlights on low beam or high beam? This was not observed. This is perhaps because there is no place for recording it on the police reporting form. Yet, it can be critical in determining why the collision happened.

In the absence of direct evidence, that is, someone reporting that they saw the headlights on high beams, I reasoned that since the collision occurred just after Mr Smith turned off a major street, where high beams would be objectionable to other drivers, that it was likely, though not certain, that the headlights were on low beam. A U.S. study looked at high beam use rates in rural, urban, and mixed rural/urban areas, and found that it was surprisingly low. Only 18% of the drivers that could, because they were separated from the vehicle ahead by 10 seconds' travel distance, used high beams (Reagan et al., 2017).

Glare

Glare from oncoming headlights can greatly reduce visibility distance. While glare from high beams has a stronger effect, glare from low beams also reduces visibility distance. Mr Smith did not see any oncoming vehicles just before he struck Mr Solomon. Thus to simulate the conditions in which the collision occurred, and in any case, for reasons of safety, there was a need for a re-enactment location where there was little or no oncoming traffic.

Streetlighting

Was there streetlighting? Yes, but the streetlights were widely spaced and the one nearest Mr Solomon was reported by the police to have been burnt out. The police also reported that the nearest operating streetlight was 40 m (131 ft) from the pedestrian when he was hit. For these reasons the streetlighting at the location where Mr Solomon was struck would not have been of much assistance in helping to illuminate a person lying on the road. That was my opinion. However, I still felt it necessary to demonstrate this was the case by carrying out a re-enactment at a site with similar streetlighting.

Consequently, the choice of a location for measuring visibility depended in large part on being able to find a section of roadway where a pedestrian could be viewed with headlights, when he or she was located 40 m (131 ft) away from the nearest operating streetlight of the same type (low pressure sodium) as the nearest streetlight present at the collision location.

The location also had to have similar surrounding lighting – in this case, that of an industrial area. Again I did not expect this commercial lighting to be a factor in visibility distance given that is generally not aimed at the roadway where it would have illuminated a pedestrian in the lane.

Traffic Volume

Last, but not least, the location for our re-enactment had to allow for visibility measures to be collected safely – it had to have little traffic. The chosen location was on a road that dead-ended at a closed industrial building, so there was almost no traffic, and what there was could be seen in plenty of time for the exemplar pedestrian "stand-in" to get off the road.

As the reader can imagine from this description of desired qualities of an exemplar location, it can be very time-consuming to find a location which meets all the criteria.

Location Selected

The location we used was a one lane per direction road in an industrial area with streetlights separated by 50 m (164 ft) on one side of the roadway.

There was a 100 m (328 ft) gap between two of the streetlights which we used to simulate the lighting at the collision site where the pedestrian had been seen 40 m (131 ft) from the nearest operating streetlight.

Re-enactment Measurements

Visibility measurements were made using a small passenger car (with similar headlight height to Mr Smith's car) on low beam headlights. Measurements were made starting more than an hour after sunset, when it was fully dark.

As at most site visits, I was accompanied by an assistant. His role was to:

- take turns with me playing the role of the pedestrian or the driver
- assist in collecting and recording the observations
- provide some security!

My assistant and I made observations of each other lying parallel to the road direction, equivalent to a north-south orientation at the accident site, as per the accident reconstructionist's report. Since the position of the hands and feet of the pedestrian were unknown, observations were made of a prone pedestrian in two positions: lying hands behind head and lying hands at side. In addition, measurements were made for the pedestrian standing on the right side of the road. This was to allow later comparison between our observations and those collected in a much larger study.

A challenging aspect of this case was that visibility measures have been recorded in a number of studies for standing pedestrians in dark clothing, but not for pedestrians lying down in the road. Because of a diminished height, they might not be seen so far away.

The pedestrian was placed 40 m (131 ft) beyond the nearest operating streetlight. The driver (myself or my assistant) drove very slowly towards the pedestrian (my assistant or myself) and stopped when the pedestrian was just detectable. We drove very slowly for two reasons. First, obviously we wanted the situation to be as safe as possible. Most of all, we could not take any chances of the person playing the role of driver not seeing the pedestrian in time. (I am always aware of the danger of repeating the original collision!) Second, during the time it takes a driver to respond to the pedestrian by pressing the brake, some distance is travelled. The distance travelled during that time should be as short as possible when the measurement is made. Otherwise the measurement distance would be overestimated. Driving very slowly minimized this error. The visibility distance was marked with a traffic cone left at the side of the road and then measured with an engineering wheel.

The last measure was made to allow us to compare our findings to visibility measures made for a large sample of standing pedestrians in other experiments, and then adjust them for a prone as compared to a vertical pedestrian. The distances for a prone pedestrian (lying with hands behind head, lying with hands at side) varied between approximately 40 and 50 m (131–164 ft).

The light levels were very low, 0.4 lux or less, on the pedestrian. To give an idea of how low that level of light is, a measurement of 1 lux is equal to the illumination of a one metre square surface that is, one metre away from a single candle. A study of seeing distance to pedestrians at a crosswalk measured visibility for two commonly used lamp types, in the presence of glare. The authors concluded that "a lighting design level of 20 vertical lux is likely adequate for proper pedestrian visibility" (Gibbons & Hankey, 2006). Thus 0.4 lux was, on these grounds, quite insufficient. Light level directly under the streetlight was measured at 8 to 10 lux. I could see my assistant at a distance of just under 100 m (328 ft) when he was standing under a lit streetlight, that is, about 2 ½ times further as compared to under a non-operating streetlight.

Above I mentioned the issue of security at site visits. Having experienced a few disturbing night-time incidents I typically bring an assistant. On one occasion I encountered (at a distance) angry parents whose son was found at fault in a boating accident. The parents were on the shore using high beams aimed so as to disrupt and invalidate a re-enactment on the water. On another occasion I was approached by a neighbour carrying a flashlight pointing at me so I could not see them, and yet another by a gentleman carrying a hockey stick, and another by curious teenagers. For this reason, I am a believer in safety in numbers, even if the number is only two!

The Science

The main human factors applying to this case were the effects of contrast, headlighting, and expectancy on seeing distance.

Contrast

The most important variable affecting visibility distance is contrast. Contrast refers to the difference in brightness between two surfaces, in this case between the pedestrian's clothing and the asphalt road surface. With the background of an asphalt roadway surface, a pedestrian wearing white clothing is more easily seen than one wearing dark clothing, because the contrast of the asphalt against the clothing is higher. A pedestrian wearing a highly reflective jacket and seen against an asphalt background would be seen even more easily.

The police report indicated that Mr Solomon was wearing jeans and a dark sweater with black shoes when he was struck. In such dark clothing, he would have been difficult to see lying on the asphalt roadway. For the re-enactment, my assistant and I wore similarly coloured clothing to Mr Solomon.

Contrast Sensitivity

When we think of measuring visual capabilities, we usually mean visual acuity, which refers to the ability of a person to resolve fine print. An equally or even more important capability is contrast sensitivity, which refers to the ability to see slight differences in brightness between a target and its background. Drivers with normal visual acuity of (20/20 or 6/6) can vary greatly in their contrast sensitivity. For example, in a study on a rural road, young subjects with 20/20 vision or better, who were passengers in a car, indicated when they could see pedestrian targets along the roadway by pressing a button. The distances at which an individual subject could see pedestrians varied greatly. For pedestrians in dark clothing standing on the right of the vehicle, the subjects with the best contrast sensitivity could see pedestrians at **five times** the distance as compared to the subjects with the poorest contrast sensitivity (Olson et al., 2010, Chapter 9: Visibility Provided by Vehicle Lighting Systems).

As drivers get older, their sensitivity to contrast decreases. As a result, older drivers have shorter visibility distances at night. This is why the ages of the individuals for whom visibility distance is measured are reported. Although it is not guaranteed that older persons will have shorter visibility distances than younger persons, it is generally the case. If the individual who was involved in a night-time collision was 75 years old for example, an investigator who is 30 years old is likely to have better visibility distance. This must be accounted for when estimating what the driver's visibility distance might have been at the time of the crash.

Streetlighting helps drivers see low contrast targets like pedestrians. However, the light varies greatly depending on where the target of interest is located. At the site visit, when the pedestrian (my assistant) stood **directly under a streetlight**, I could see him at 2 ½ times the distance compared to a location 40 m (131 ft) from the nearest operating streetlight.

Visibility with Headlights Alone

So, in this case where the nearest streetlight is burnt out and the remainder of the streetlights do little to illuminate the area in which the pedestrian lay, how far can you see a pedestrian with low beam headlights alone? The experiment at the Highway Safety Research Institute at the University of Michigan referred to above provides some answers. Researchers measured the ability of younger and older subjects to see various kinds of targets, including pedestrians dressed in light and dark clothing, in a dark rural area, to the left and to the right of the road, using low beam headlights.

Subjects were driven towards the pedestrian targets. As soon as they saw a target, they responded by hitting a button. When pedestrians in dark clothing stood on the right side of the road, the median visibility distance at which the 23 young subjects hit the button was 40 m (131 ft). Older subjects (aged 60–75) years responded at half that distance.

A median distance means that half of the subject sample had visibility distances better than 40 m (131 ft), and half, less than 40 m. That is, half of the subjects could see the pedestrian further than 40 m, whereas the other half saw the pedestrian at 40 m or less.

The visibility distances reported here are actually "response distances". (I once got into big trouble for not distinguishing between visibility distances and response distances when giving evidence in court. More on this in Chapter 12.) Response distances have to be adjusted to account for the time it takes for the subject to put their foot on the brake after perceiving the pedestrian lying on the roadway. The faster the subject is travelling the more distance is going to be eaten up during perception-reaction time. Consequently, visibility distance will be underestimated. For the purposes of accurately measuring visibility, subjects in this experiment were driven at a slow speed, of 40 km/h or 11 m/sec. The button press took about half a second, during which time, the subject travelled about 6 m. Thus, the subjects' median visibility for pedestrians standing on the right was 40 m plus 6 m or 46 m based on the distances measured for pedestrians in dark clothing, positioned on the right side and the driver travelling at 40 km/h (Olson et al., 2010).

Our estimates of the visibility of a pedestrian standing on the right based on our observations during the site visit were similar – 44 and 49 m (141 ft and 160 ft) as compared to 40 m (131 ft).

Visibility of Prone vs Vertical Pedestrian

As noted earlier a challenging aspect of this case was that while visibility measures have been recorded in a number of studies for standing pedestrians in dark clothing, we did not know of studies involving pedestrians lying down in the road. Because of a diminished height, they might not be seen so far away. Consequently, a correction for a prone versus a standing pedestrian was required. During our site visit we found that, based on our observations, a pedestrian lying down would be seen at a distance 8% less than a standing pedestrian.

After correcting for the diminished height (8% reduction), the estimated median visibility distance, based on Olson's numbers, of 46 m (151 ft) became 42 m (138 ft).

Low beam headlights are aimed down and to the right to avoid glare for oncoming drivers and as a result visibility distances are longer on the right they are on the left. Mr Solomon was seen in the centre of the roadway where distances would be similar to those on the right. In Olson's experiment, the distances at which subjects detected the targets in dark clothing standing on the right ranged from 21 to 82 m (69 ft to 269 ft) with a median of 46 m (151 ft). These are 5th–95th percentile responses. In other words, a total of 95% of the population had visibility equal to or exceeding 21 m (69–269 ft). Only 5% had visibility less than 21 m (69 ft) the 5th percentile response. On average older drivers responded at shorter distances.

Expectancy

An important factor that must be considered in determining visibility distance is driver expectation. A pedestrian lying in the middle of a roadway at night would be a most unusual and unexpected sight. Whether or not a driver is expecting to see something in the road has a great impact on visibility distance. Knowing the nature of the target makes it possible to use subtle cues, such as white treads on the pedestrian's shoes, to detect them further away than might be possible under normal conditions. Those values measured in experiments where the subject knows there are targets to detect, and what those targets look like, are going to be much higher than those in the real world.

The difference expectation can make in the detection of targets at night is considerable. It was first illustrated by a study of pedestrian visibility performed in the United States in the late 1930s. In this experiment subject drivers were told that they were to assess the quality of the headlights on the experimental vehicle. They drove around for a while with the experimenter. Then they gave an assessment of the headlights and were told to go back to the lab to get paid. On the way back to the lab, the real experiment happened. They encountered a pedestrian target in the roadway, without any forewarning. The car was instrumented so that distance and time at which the driver started to release the accelerator could be measured. Once the driver knew exactly where the target was and what it looked like, the experimenter then asked the driver to back up the car and drive forward again at the same speed as on the first encounter. Then the distance at which the subject could recognize the target was measured again. On average, response distances when the target was unexpected were half of those obtained where the driver knew where the target was and was expecting it (Roper and Howard, 1938). This is a very important result because it brings into question evidence on visibility distance occasionally still given in court based on a re-enactment. For example, the investigator says he could see the target at a given distance, say 60 m. He concludes that the driver involved in the crash also ought to have been able to see the target at that distance. And having done so would have been able to stop in time. However, if expectancy were to be considered, a more realistic visibility distance would be half that, or 60 m, not enough for the driver, depending on his or her speed to have stopped.

The large effect of expectancy has since been substantiated (Shinar, 1985), although it is not possible to "surprise" subjects as was done so thoroughly in the 1938 study described above, due to modern ethics codes regarding the treatment of subjects. Based on these studies, Dr Paul Olson, a U.S. expert in visibility at night with headlights, suggests taking the median visibility distance derived from test data with alerted subjects and reducing it by half to approximate a more realistic estimate of the median response possible under conditions of normal expectancy.

Based on the data from Olson's experiment, our estimate of the median visibility distance, under ideal conditions (cleaned and freshly aimed headlights, driver expecting target), when corrected for the diminished height of the pedestrian, was 42 m (138 ft). We corrected for expectancy by cutting the estimate by half. Our estimate of the median visibility distance was then 21 m (69 ft).

In other words, half of drivers would have a visibility distance of 21 m (69 ft) or less in the situation where there was no streetlighting, the pedestrian was lying in the roadway and drivers were not expecting to see them.

Stopping Distance

The last piece of the puzzle we need to consider is stopping distance. That is, once a driver's foot hits the brakes how far does the driver travel during braking before stopping? This calculation is generally the province of the crash reconstructionist. This distance depends on a number of factors such as speed, weather conditions, pavement friction, condition of the brakes, type of brakes, weight of the vehicle, etc. It also depends on how quickly the vehicle is braked. In an emergency situation such as this, a high level of deceleration is assumed. With a speed of 50 km/h (30 mph) and assuming normal pavement and weather conditions, and perception-reaction time of 1 second, emergency stopping distance is approximately 30 m (100 ft), longer than the available visibility distance of 21 m (69 ft) or less for half of drivers. Perception-reaction time is short (1 second) as expectancy has already been accounted for. Thus, a majority of drivers could not stop in time (Olson Chapter 13 in Smiley, 2016).

We can clearly see that visibility at night with low beam headlights is not very good, especially if one has to consider age, glare and expectancy among other things. Pedestrians do not think so however! Pedestrians walking on the shoulder of a closed road circuit were asked to indicate when they were confident that an approaching driver could first recognize that they were present. The results of the study were that pedestrians overrate their visibility distance by a **factor of two** (Tyrrell et al., 2004). At night it really is best to consider oneself as invisible to oncoming drivers when crossing a roadway.

Conclusions

In summary, there are a number of factors that contributed to the poor visibility of Mr Solomon. It was fully dark. Mr Solomon was lying prone in the roadway, wearing dark clothing and totally unexpected by the driver. He was most likely seen with low beam headlights. There were streetlights, but the nearest functioning streetlight was 40 m (131 ft) away and did not provide enough light for a driver to detect Mr Solomon in time to avoid striking him.

Based on experimental results from a large sample of young and old drivers, adjusted for pedestrian position, half of drivers would see an unexpected prone pedestrian at a distance of 21 m (69 ft) or less. However, half of drivers would require 30 m (100 ft) or more to come to a full stop from 50 km/h (30 mph). I concluded that only a minority of drivers, less than half, would have been able to detect Mr Solomon in time to come to a full stop from a speed of 50 km/h (30 mph) using emergency braking before striking Mr Solomon.

In my opinion, by lying in the road in dark clothing at night, Mr Solomon put Mr Smith in a challenging position that only a minority of drivers could handle successfully.

Other Human Factors

As indicated in the introduction to this book I have not considered all the factors at play in each collision. There are so many human factors, one could write a book about each case! Mr Solomon had been drinking heavily and used poor judgement in choosing to lie down on the roadway. Drinking is even more risky for pedestrians than it is for drivers. Of the pedestrians involved in traffic crashes that resulted in pedestrian deaths, 35% had a blood alcohol concentration (BAC) of 0.08 g/dL or higher, the legal limit in many jurisdictions. Less than half of drivers had a BAC that high (NHSTA, 2007).

Countermeasures

Federal Highway Administration (FHWA) statistics indicate that three quarters of pedestrian fatalities in the United States occur at night. The obvious countermeasure in the case of this collision would be to improve streetlighting. However, safety costs money and a site so isolated as is the subject of this chapter, and with few pedestrians and minimal traffic (a dead-end) would not merit consideration for such. Another way of improving the lighting situation is to address the substantial under use of high beam headlights. A U.S. study looked at high beam use rates in rural, urban and mixed rural/urban areas, and found that it was surprisingly low. Only 18% of the drivers that could, because they were separated from the vehicle ahead by 10 seconds' travel distance, used high beams (Reagan et al., 2017).

Advanced technologies such as high beam assist, which switches automatically between high and low beam headlamps depending on the presence of other traffic, can help to address this problem.

References

Gibbons, J. M., & Hankey, R. B. (2006). Influence of vertical illuminance on pedestrian visibility in crosswalks. *Transportation Research Record, 1973*(1), 105–112.

National Highway Traffic Safety Administration. (2009). Traffic safety facts: 2007 data: Pedestrians. *Annals of Emergency Medicine, 53*, 824.

Olson, P. L. (2016). Visibility provided by vehicle lighting systems. In A. Smiley (Ed.), *Human factors in traffic safety* (p. 323, 339). Lawyers & Judges Publishing Company.

Olson, P. L., Dewar, R. E., & Farber, E. (2010). Visibility provided by vehicle lighting systems. In D. Krauss & P. Olson (Eds.), *Forensic aspects of driver perception and response* (3rd ed., p. 206). Lawyers & Judges Publishing Company.

Reagan, I. J., Brumbelow, M. L., Flannagan, M. J., & Sullivan, J. M. (2017). High beam headlamp use rates: Effects of rurality, proximity of other traffic, and roadway curvature. *Traffic Injury Prevention, 18*(7), 716–723.

Roper, V. J., & Howard, E. A. (1938). Seeing with motorcar headlamps. *Illumination Engineering, 33*, 412–438.

Shinar, D. (1985). The effects of expectancy, clothing reflectance, and detection criterion on nighttime pedestrian visibility. *Human Factors, 27*(3), 327–333.

Skyving, M., Möller, J., & Laflamme, L. (2023). What triggers road traffic fatalities among older adult drivers? An investigation based on the Swedish register for in-depth studies of fatal crashes. *Accident Analysis & Prevention, 190*, 107149.

Tyrrell, R. A., Wood, J. M., & Carberry, T. P. (2004). On-road measures of pedestrians' estimates of their own nighttime conspicuity. *Journal of Safety Research, 35*(5), 483–490.

3

Red-Light Runner vs Driver

Introduction

An inattentive driver ran a red light. His vehicle was struck by another driver crossing the intersection. The main human factors issue explored in The Science section deals with the ability of a crossing driver who has a green light to detect a red-light runner in time to do anything about it.

The Collision

One fine spring morning, a young woman, Christine Brown, was on her way to work, driving as she normally did along Minor Road. As she approached the intersection with Major Avenue, her traffic light turned from green to caution and then to red. She stopped at the light. To her right, several pedestrians and bicyclists were crossing or waiting to cross.

At right angles to her, on Major Road, Dr Ford was approaching his green light in no particular hurry. As Dr Ford neared the intersection, the green light turned amber and then red. Dr Ford's speed did not change. He continued through on a red light.

When the light turned green for the drivers on Minor Road, one pedestrian was still crossing in front of Ms Brown. As he was crossing, and she was stopped, waiting, she remembers looking left and looking right for traffic. As Ms Brown did so, she noticed a nanny pushing a double stroller diagonally across the intersection. Ms Brown noticed because the design of the double stroller was of interest to her at the time.

Once the last pedestrian had finished crossing in front of her, Ms Brown pulled out, looking ahead of her across the intersection to the continuation of Minor Road. After she entered the intersection, Dr Ford, failing to notice he had a red light, also entered the intersection. Ms Brown struck the left side of Dr Ford's vehicle. The force of the collision caused Ms Brown's sport utility to spin around into the nanny and her two charges, striking and seriously

DOI: 10.1201/9781003564805-3

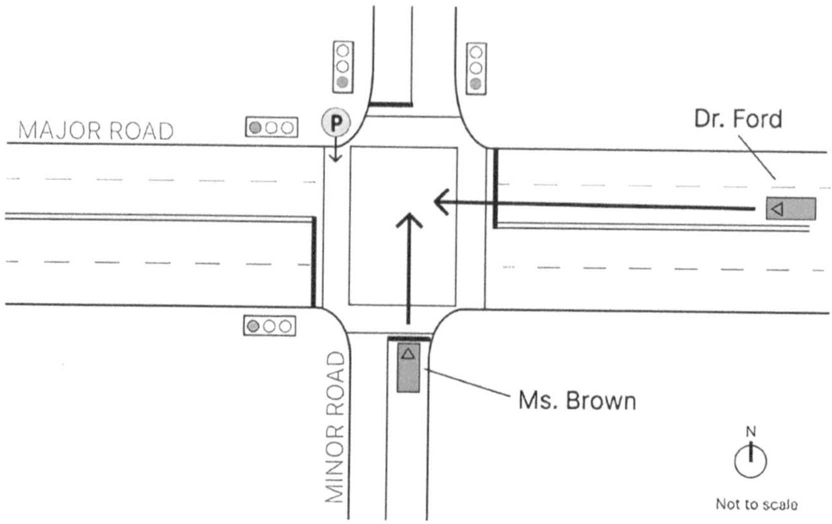

FIGURE 3.1
Collision diagram showing the location of involved parties.

injuring them. Ms Brown was also injured. Dr Ford was shaken but not injured.

The collision diagram in Figure 3.1 shows the paths of the two vehicles and the path of the nanny with the double stroller (P) a few seconds before the crash.

The Mandate

Dr Ford was charged with failure to obey a red signal. Dr Ford's family sued Ms Brown, claiming that, although Dr Ford was at fault, she carried part of the blame. This is because, they claimed, she had the opportunity to avoid the collision had she looked to her right before pulling into the intersection. The aim of my human factors investigation was to see how easy or difficult it would be for a driver in Ms Brown's position to detect a red-light running vehicle in time to avoid a collision.

It always seems strange to me that in collisions in which one person is clearly at fault, the not-at-fault driver is routinely sued for what they "could have" done to lessen the damage or injury. This is because civil lawsuits involve a process through which one person can seek to hold another person liable, that is, sue, for some type of harm or wrongful act. Liability is seldom 100% for one party and 0% for the other, no matter how much at fault the first party is. A human factors assessment is helpful for determining how each party might have contributed to the outcome.

The Investigation

I visited the collision site with my assistant under traffic and lighting conditions like those at the time of the collision. The collision had occurred during the day and at an off-peak traffic time, so that is when we visited.

The aim of our investigation was to determine why Ms Brown failed to notice the car that ran the red light in time to stop and avoid a crash. First, we needed to consider where Ms Brown would likely have been looking in order to carry out her task of driving through the intersection safely.

During the site investigation, we follow the path of the involved drivers to the extent that it can be done safely. We typically record video from the driver's perspective using a mini camera mounted on a pair of safety glasses. The video shows where the driver's head is pointing, which is generally where the driver is looking or close to it. The safety glasses do not interfere with a driver's vision any more than a pair of sunglasses would. Figure 3.2 shows the driver wearing glasses with a mini camera. Figure 3.3 shows a frame grab from a video as an example of the view we record.

By observing the intersection of Major and Minor Road from Ms Brown's point of view, we saw that the intersection was a busy one. It was near a college. There was a steady stream of bicyclists and pedestrians during the morning on a weekday, which is when the collision occurred.

By following her path, we could see that just before and just after the light changed, Ms Brown would need to keep an eye on:

- the traffic signal which was about to change
- crossing pedestrians and bicyclists still in the crosswalk
- the sidewalks, to make sure no more pedestrians were going to step onto the road against a red light
- right-turning drivers across the intersection with whom she would have to merge as they turned onto Minor Road
- the path through the intersection that she was going to follow

She told the police that Minor Road was narrow for her because of the many parked cars and her wide sport utility vehicle. Consequently, she was always careful around bicyclists on that street. Clearly, there was a lot to pay attention to in this intersection.

Ms Brown said that she had looked left and right for traffic before starting to move forward. In particular, she remembers looking at the slow pedestrian

FIGURE 3.2
Driver wearing safety glasses with mini camera embedded between eyes.

FIGURE 3.3
Frame grab from video recorded by mini cam worn between driver's eyes.

who was walking across the street directly in front of her vehicle. She remembers noticing the double stroller when she looked left. We assumed from her evidence that she did indeed also look right in the direction of Dr Ford. If she did notice Dr Ford approaching in the distance, at what distance from the stop bar (painted white line where drivers are to stop) could she tell that he was not going to stop? Would she have enough time to stop before crossing Dr Ford's path?

To help to answer these questions, we first identified the point where Dr Ford and Ms Brown collided. This is called the point of impact and can often be found in the Motor Vehicle Collision Report completed by the police. The point of impact is identified in various ways, including, but not limited to, brake or skid marks or property damage. Then we watched traffic travelling in the same direction (north), as Dr Ford did, as he approached the red light at Minor Road. We wanted to see how far from the stop bar at the intersection we could tell that a vehicle was starting to slow.

Our measuring device was approximate. We used the number of lane marks as a measure of distance (see dashed lane marks in Figure 3.1). Lane marks are the white stripes used to divide adjacent lanes going in the same direction. We first could detect slowing when vehicles were two to three lane marks south of the stop bar and three to four lane marks south of the point of impact where Dr Ford and Ms Brown collided (see Figure 3.4).

The lane mark and the gap to the following lane mark were measured to be 10 m (33 ft). Two to three lane marks is approximately the distance of 20 to 30 m (66–98 ft). In other words, northbound vehicles can get quite close to the stop bar before it is obvious that they are **not** going to stop. The point of impact was about 10 m (33 ft) beyond the stop bar.

FIGURE 3.4
View of Ms Brown to right when she is stopped and approaching vehicle at lane mark 3.

The distance from where westbound vehicles (i.e. the direction that Dr Ford was travelling) can be observed to start slowing to the point of impact within the intersection was measured to be 30–40 m (98–131 ft). It would have taken Dr Ford only 2–3 seconds to cover this distance at an approach speed of approximately 50 km/h (30 mph).

At the site we recorded the times taken for vehicles following the path of Ms Brown to leave the stop bar and reach the point of impact identified by the police. The times recorded ranged from 3 to 4 seconds.

In other words, Ms Brown would not have been able to detect that Dr Ford was not slowing until a second or two after she left the stop bar. At this point, it was too late. Her attention would have been on the narrow path in front of her on the continuation of Minor Street and on the pedestrians and bicyclists around her. Even if she did notice Dr Ford enter the intersection, it would likely be too late to brake and avoid a crash. Furthermore, she would also not expect someone to drive through a red light so long after the traffic light had changed.

The Science

The scientific issues which were important in this case concern driver looking behaviour and distraction. Would most reasonably attentive drivers have **noticed** the red-light runner, Dr Ford, in time to stop before striking him? Or was Ms Brown inattentive (perhaps looking at something too long, like the double stroller) given the driving task at hand?

At a busy intersection, there are many places at which to look – the road ahead, the traffic signal, corners where pedestrians may be about to cross, and so on. All these places cannot be looked at at the same time. What does the driver notice? How is that affected by the design of the eye? How is that affected by the difficulty of the driving task?

The Design of the Eye

Throughout this chapter, I will be putting the words "**see**" and "**notice**" in bold letters. This is to emphasize that they are different. You are capable of **seeing** many things from a given point of view. But you can **notice** only a small portion of these. What you **notice** depends not only on the design of the eye. It also depends on how the brain processes information.

The visual field of the human eye is large, about 90° to the left and 90° to the right of the line of sight. This is the extent of the visual field that is relevant to an optometrist but not as much to someone interested in driver search behaviour (see Figure 3.5).

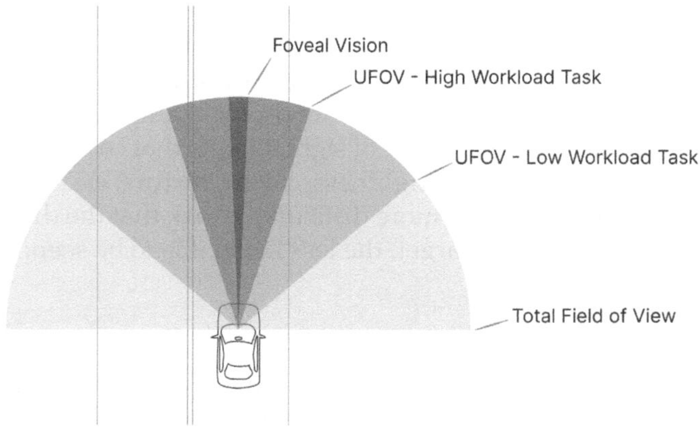

FIGURE 3.5
Eyes' field of view. Foveal Vision 2-4 degrees, Total Field of View 180 degrees.

With effort, and without turning your head, you can see a target many degrees off the line of sight while using peripheral vision. But it is hard work, and you would be most unlikely to notice the target unless your attention was drawn to it and you had lots of time to look at it. Two different ways of having attention drawn to a target are called top-down and bottom-up. I will get to how they affect search in a minute. But first, let's think about the design of the eye.

Foveal Vision

While the eye's field of view is large, only a small area of the retina (equivalent to film in a camera), at the back of the eye, is high resolution. High resolution means you can see details easily. This small high-resolution area is called the fovea and covers a cone of about 2–4° (see Figure 3.5). This area is packed with individual light sensors called rods and cones. If you hold a quarter up at arm's length, the dimensions of the quarter are the approximate size of the high-resolution area of your eye. When you search a scene, the road, for example, you keep moving your eyes rapidly from one target (e.g. the traffic signal) to the next (e.g. the pedestrian), each time repositioning your "quarter" to get the best view. Anything seen outside that quarter is seen in peripheral vision. Due to the way the eye is designed, drivers search the road scene in a series of rapid fixations. A fixation refers to the short period of time the eye rests on a target.

Peripheral Vision

There is **not** a hard boundary between foveal and peripheral vision. The further off the line of sight an image is seen in peripheral vision, the fewer light sensors are stimulated and the fuzzier the image. Try reading newspaper headlines with your peripheral vision. Even though headlines are very large, it is difficult to read them unless you are looking directly at them with your high-resolution foveal vision. The things we **notice** in our low-resolution peripheral vision are areas with many targets, for example, cars, signs, pedestrians, targets that are near our line of sight, targets that contrast greatly with their backgrounds in terms of brightness, colour, texture, etc., targets of large size, and that are moving. The more difficult the task that the driver is doing, and the more unexpected a target, the less likely it is to be seen.

Useful Field of View (UFOV)

Between foveal vision (2–4° centred on the line of sight) and the maximum extent of the field of view (approximately 90° to the left and 90° to the right or 180°) is an area called the useful field of view (UFOV). Figure 3.5 shows the three areas in the field of view. UFOV is defined as the total visual field in which useful information can be acquired in a single glance without eye and head movements. Your UFOV may be more or less restricted depending on the mental load associated with what you are doing (Wolfe et al., 2017). For example, driving on a residential road with no other traffic would be considered low workload, whereas driving on the freeway in rush hour would be considered a high workload task. A given target is more likely to be noticed in a low workload situation. (The UFOV has mainly been studied in relation to predicting older driver crash rates, with a general finding that drivers with higher crash rates have smaller UFOVs.)

Target Detection and Mental Load

On-the-road studies have looked at the target properties and driving circumstances which affect target detection.

In an Australian study, researchers set up round disc targets, mounted on poles like traffic signs along several stretches of roadway. The targets had selected sizes and distances off the roadway, and were in different driving environments.

A first group of drivers was asked to report whatever objects they **noticed** as they drove (e.g. traffic sign, woman standing at bus stop, mailbox, etc.) (Hughes & Cole, 1984). In other words, driver subjects were asked to report their stream of consciousness while driving. Subjects were unaware that the researchers were most interested in the round disc target signs.

A second group of driver subjects were given different instructions. They were told, in advance of starting to drive, about the disc targets that had been set up along the roadway. This group of subjects was asked to search for the disc targets and report them.

The drivers (second group) who were asked to search for the disc targets specifically were considerably more likely to notice them (40% of targets detected) and report them as compared to the subjects (first group) who were not asked to search for them specifically but just to report whatever they saw (6% of targets detected). This is a large difference (40% versus 6%) and shows the importance of expectation in target detection. The more you expect to see something, the more likely you are to detect it.

Subjects drove in different driving environments. When they drove in a busy urban area with shops, far fewer of the disc targets were reported (40% of targets detected) compared to driving in a peaceful residential area which was much less visually busy (80% of targets detected). This is also a big difference and shows how important mental workload is in how well drivers detect information in the driving environment. The shopping area would have had many more things to look at than the residential area, and not surprisingly more disc targets were missed in the shopping area.

Target Detection and Eccentricity

In the study just described, another important finding was that objects tended to first be **noticed** when they were near the line of sight, that is, at small angular eccentricities (when they were less than 10° off the driver's line of sight).

In a second study detection time was measured for lights placed within the vehicle at various angles off the driver's forward line of sight, that is at various eccentricities, while drivers drove in various environments, highway, rural road, and urban streets. Drivers responded to the appearance of the light illuminating by touching a metal ring mounted on the steering wheel. If a light was not responded to within 4 seconds, it was turned off and counted as a miss. Subjects in the second study were chosen from among well-experienced police car drivers (Cohen, 1987). On urban streets, the busiest and most cluttered environment, these drivers missed more than a quarter of the target lights when the angle between the target and the driver's line of sight was 15°. When the angle increased to 25° about 50% of targets were missed. Perception-reaction time for targets 15° off the line of sight was much longer than for targets seen up to 5° off the line of sight.

A third study of driver visual search was done for the purpose of developing policy related to video advertising near the roadway. Drivers travelled around the downtown of a major city through major intersections with video signs, wearing an eye tracker (Smiley et al., 2004). An eye tracker is a very accurate way of measuring the length and position of eye fixations. The eye tracker does not interfere with the driver's view any more than a pair of sunglasses. A beam of infrared light is bounced off the cornea, and as the eye moves the angle of reflection changes. These changes in angle of reflection are recorded by the eye tracker. In addition, the road scene is recorded. These data are used to produce the eye position on the related video image. Figure 3.6 shows a screen grab from an eye tracker. The white cross (highlighted to improve contrast) indicates where the driver is looking.

FIGURE 3.6
Frame grab from eye tracker video.

Despite the attention-getting nature of the signs, that is, positioned close to the driver's line of the line of sight, large, moving, colourful, etc. drivers did not look very far off the line of sight along the roadway. As has been found in other studies the vast majority (three quarters) of eye fixations are at the road ahead (Smiley et al., 2004). A related master's thesis on the distracting effect of advertising signs, also used an eye tracker to collect data on how far off the line-of-sight drivers typically first looked at (**noticed**), advertising signs (Beijer 2002).

The studies described above indicate that the UFOV while driving is in the range 10–15° off the driver's line of sight. That is where most first looks (**noticing**) will occur.

The studies described above show the majority of driver fixations are on the road and roadside ahead, a strategy that allows targets to be picked up on both the left and the right in peripheral vision. Fixations last from 0.25 to 1.5 seconds (Olson et al., 2010, p. 86).

Top-Down versus Bottom-Up Search

Visual search can be done in a top-down manner. This means that the brain consciously selects where to look depending on the driving task (e.g. look left and right for cross traffic just before pulling into an intersection). This was how the second group in the Australian sign disc study were instructed to search (look for and report disc targets) (Tijerina, 2016).

Visual search is also bottom up, that is, determined by what is **noticed** in peripheral vision. And that will be determined by such characteristics as size, contrast, eccentricity and so on qualities that are most attention-getting in peripheral vision. This was how the first group in the Australian sign disc study were instructed to search. (Report whatever you **notice.**)

Analysis

Did Ms Brown use top-down search? According to her evidence, she checked for traffic to the left and to the right just before leaving the stop bar, a conscious decision.

This is because, as shown in Figure 3.7, a few seconds before the collision there was a 70° angle between the vehicles, and thus Dr. Ford would be seen at 70°off Ms. Brown's line of sight. The 70° angle is determined by the relative speeds and distance travelled of the two colliding vehicles as they approach impact. The greater the difference in speeds, the greater is the difference in distance travelled by Dr. Ford as compared to Ms Brown, and the more off her line of sight he would be seen, and the less likely he would be to be noticed by her. See Figure 3.7 with eccentricity of 70°.

In regard to the ease with which the Ford vehicle could be detected in peripheral vision, the evidence from one of the pedestrians is of interest. He said that he had started crossing without looking right, towards the car that ran the red light. He did not look because he thought he could see out of

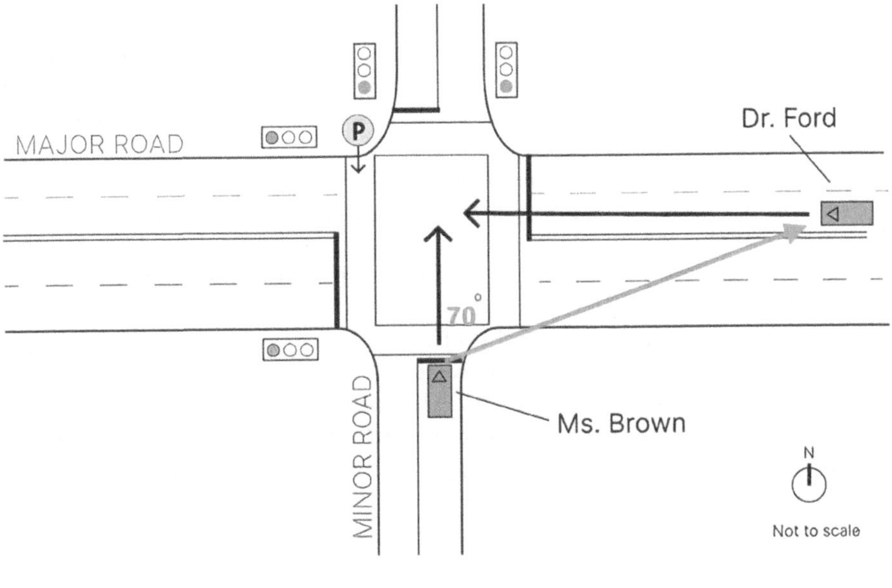

FIGURE 3.7
Collision diagram showing the location of involved parties.

the "corner of his eye" that nothing was coming. So, it was completely unexpected when he (Dr Ford) blew the red light. Like Ms Brown, he did not detect the Ford vehicle in peripheral vision in advance of it running the red light.

Ms Brown did not have any particular reason to expect in advance that there was a car coming towards her from her right about to run a red light (i.e. requiring top-down search to the right). Nor would she have expected a driver to run the red so long after the light turned red. There was a 2 seconds all red interval. There was evidence from the pedestrian about where he was and his likely walking speed when Dr Ford went through the intersection. This evidence indicated that in total the delay between the light turning red and Dr Ford going through the intersection was about 6 seconds. Most run-the-red light offences occur during the first couple of seconds – not 6 seconds later.

Studies using eye trackers show that the number of fixations that can be made, and the number of objects that can be identified as a driver drives through a particular area is very limited. The best visual strategy for Ms Brown would have been to turn her head and look to the left and to the right before starting to accelerate into the intersection, which she said she did, and then to look directly ahead as she moved forward. This would allow detection of slow-moving pedestrians, bicyclists and cars in peripheral vision on either side near her vehicle. Once she started moving forward, Ms Brown's primary task was to look ahead at the path along Minor Road. The path was narrow because of the wide vehicle she was driving and the parked vehicles.

Unfortunately, this optimum strategy would not enable Ms Brown to detect Dr Ford's vehicle in time to avoid a collision. As we discovered during the investigation, we first could detect slowing that came before a stop when vehicles coming up to the intersection were quite close, two to three lane marks south of the stop bar. It would take Dr Ford only 2–3 seconds at the speed limit to cover this distance, not enough time for Ms Brown to realize that she was not going to stop and bring her vehicle to a stop to avoid running into him.

In this case, Dr Ford's car would have been **seen** at an angle of about 70° by a driver in Ms Brown's position, looking straight ahead. Most drivers travelling straight through an intersection can be observed to be looking straight ahead while crossing it. An angle of 70° is considerably larger than the 10° to 15° off the line-of-sight threshold to each side (total 20–30°) within which most targets are **noticed** in peripheral vision.

Not only that, suppose she had detected Dr Ford's car in peripheral vision, there would not have been time for Ms Brown to carry out the necessary tasks to avoid it. These tasks included **noticing** his vehicle in peripheral vision approaching the intersection, glancing at his vehicle to **see** it clearly enough (foveal vision) to realize it was not going to stop, hitting the brake, and stopping her vehicle before the collision. After all, she was already in the intersection when he entered it.

Measurements we made at the site show it would have taken Ms Brown only 3 to 4 seconds from when she left the stop bar to reach the point of impact with Dr Ford's vehicle in the centre of the intersection. During this time the number of additional glances left and right that could have been made, beyond the stop bar, requiring about 1–2 seconds each, would have been very limited.

Distraction

Ms Brown faced, what was for her, a distracting scene of the nanny and her two charges, diagonally, across the intersection, to her left. Given its location, to her left, the distraction could have taken Ms Brown's attention away from her right. This was important because Dr Ford was approaching from the right. It may have lengthened any glances to the left, resulting in less time looking towards the right.

Conclusions

Thinking about where Ms Brown had to look and for how long as she crossed the intersection, is it reasonable to have expected her to detect Dr Ford, realize he was going to run the red light, and brake in time to prevent a collision?

In my opinion, the answer is "no". First it is clear that because of the large angle between her line of sight and Dr Ford's vehicle, she would have been unlikely to **notice** it out of the corner of her eye, that is in peripheral vision, as she started to cross the intersection. She would have had to look in his direction away from the narrow path ahead of her. Had she done so, it would take her at least a second to assess the Ford vehicle distance and speed and determine whether to hit the accelerator or the brake. By the time it was possible to detect that Dr Ford was not going to slow (as little as 2 seconds from impact), Ms Brown would already have left the stop bar and entered the intersection. Furthermore, she would have been concentrating on the path ahead directly in front of her to Minor Street on the other side of the intersection and on the pedestrians and bicyclists immediately around her. She may also have been distracted by the nanny with the double stroller, causing her to look longer in the direction opposite that of the approaching red-light runner, Dr Ford.

The Courtroom

This case went to trial. There was an opposing human factors expert with whom I was very familiar. His opinion was that Ms Brown should have seen the red-light runner. It is always a bit awkward when experts disagree, and furthermore, when one knows the opposing expert! In this case the judge preferred my analysis of the collision. As usually happens with court cases, I never learned the portion of liability assigned to each participant in the crash.

Countermeasures

There are many countermeasures that have been demonstrated to reliably change driver behaviour and reduce intersection crashes, such as by adjusting the timing of the amber (caution) traffic light, or by installing red-light cameras. However, neither of these would likely have changed the behaviour of the red-light runner in this case. What might have worked instead this case is a collision warning system that specifically deals with driver red light running at intersections. These are still being developed and tested (Chang et al., 2009) and it is important for drivers to remember that not all collisions can be warned about.

References

Beijer, D. D. (2002). *Driver distraction due to roadside advertising* (Master's Thesis). University of Toronto, Department of Mechanical and Industrial Engineering.

Chang, U. S. H., Lin, C. Y., Hsu, C. C., Fung, C. P., & Hwang, J. R. (2009). The effect of a collision warning system on the driving performance of young drivers at intersections. *Transportation Research Part F: Traffic Psychology and Behaviour, 12*(5), 371–380.

Cohen, A. S. (1987). The latency of simple reaction time on highways: A field study. *Public Health Reviews, 15*, 291–310.

Cole, L., & Hughes, P. K. (1984). A field trial of attention and search conspicuity. *Human Factors, 26*(3), 299–313.

Olson, P. L., Dewar, R., Farber, E. (2010). *Forensic aspects of driver perception and response* (3rd ed.). Lawyers & Judges Publishing Company, Inc.

Smiley, A., Smahel, T., & Eizenman, M. (2004). Impact of video advertising on driver fixation patterns. *Transportation Research Record, 1899*, 76–83.

Tijerina, L. (2016). Information processing, attention, and driver workload. In A. Smiley (Ed.), *Human factors and traffic safety* (Section 3.3: Expectancies and top-down vs bottom-up processing) (p. 3). Lawyers & Judges Publishing Company.

Wolfe, B., Dobres, J., Rosenholtz, R., & Reimer, B. (2017). More than the useful field: Considering peripheral vision in driving. *Applied Ergonomics, 65*, 316–325.

4

Motorcycle vs Train

Introduction

This chapter describes a collision involving a motorcyclist and a train moving through a crossing at night. The motorcyclist failed to see a train in the crossing until it was too late to stop. This rail crossing had only a simple railway crossing warning sign. In 2015 in the United States, 55% of crossings had active warning systems, that is, flashing lights or gates to provide real-time warnings: 45% had only a passive warning sign (Gabree et al, 2019). The main human factors issues discussed in The Science section are the visibility of a train at night, and the limitations of warning signs at train crossings. Solutions to better warn drivers of the presence of a train at night are described. The motorcyclist successfully sued the railway for not providing adequate visibility of the train.

The Collision

This chapter describes a collision involving a motorcyclist and a train at a crossing in a rural area at night. As the motorcyclist drove towards it, he failed to see a train in the crossing until it was too late to stop. He passed between two black oil tanker cars and ended up on the other side of the train. He survived, albeit with serious injuries. This rail crossing was the most basic of crossings. There was just a simple railway crossing sign to warn drivers that there was a railway crossing at this location. The main human factors issues discussed in The Science section are the visibility of a train at night, and the limitations of warning signs at train crossings. Solutions to better warn drivers of the presence of a train at night are described. The motorcyclist sued the railway for not providing sufficient visibility of the train. The case went to trial. The judge's findings are discussed.

DOI: 10.1201/9781003564805-4

A young man on a motorcycle, Jason Zsoldos, rode down a gravel road towards his home in the country. His route took him across a single-track railroad crossing deep in a rural area. This crossing was the most basic of crossings. When a train approaches, there are no gates that come down and there are no flashing lights or bells at the crossing. There is just a simple railway crossing sign to tell drivers that there is a railway crossing at this location (see Figure 4.1). It was late at night, and it was overcast. There were trees and bushes framing the crossing. There was no moon, so it was very dark. The only light was from the motorcycle headlight.

According to Mr Zsoldos, he rode across the tracks regularly, including at night. He seldom saw trains, and especially seldom at night. Some time before Mr Zsoldos arrived at the collision point, a long train had entered the crossing. The front engine was 35 cars ahead of the crossing or about 30 seconds' travel, and so its headlight would not have been any help to him in warning of the train. Mr Zsoldos hit black oil tankers. The railcars had little in the way of reflectors or reflective striping (see Figure 4.2).

He saw the train too late to stop and shot through between two tankers, ending up on the far side of the tracks severely injured. As a result of his collision with the train, he had to have both arms and one leg amputated. Amazingly, in this isolated area, late at night, someone happened upon him

FIGURE 4.1
Unprotected rail crossing – warning sign only.

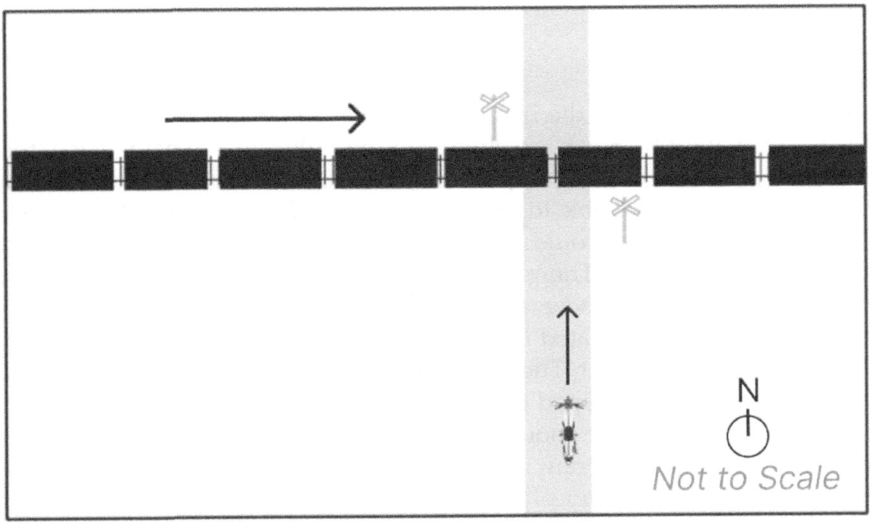

FIGURE 4.2
Collision diagram

minutes after the collision. This good Samaritan took him to hospital. He survived.

Mr Zsoldos had been drinking, but because of loss of his arms and leg, and blood loss, the grim conclusion was that it was difficult to tell how much he had been drinking, and therefore whether alcohol had contributed or not to this collision.

The Mandate

The aim of the investigation was to assess visibility, that is, the seeing distance to the train for the majority of reasonably alert, unimpaired drivers, and to see whether Mr Zsoldos should have been able to see the train in time to stop or was prevented from doing so for some other reason, such as inattention, riding too fast, or by the effects of alcohol.

According to the lawyer for the motorcyclist, Mr James Mays, the central concern of the case from a legal point of view would be phrased "would a train occupying this crossing pose a hazard to a driver with average capabilities" (referred to in law as a "reasonable" driver). "Average capabilities" is not intended to refer to a statistical average but rather average in the sense of "commonplace".

The Investigation

The challenging part of the investigation was to be able to view a train in the crossing at night. Since the owners of the railway were being sued, they were unlikely to be willing to provide times at which trains would use the crossing. Nor were we able to identify a nearby store, restaurant, or gas station whose employees could help us with likely crossing times. And so, one night my colleague Thomas Smahel and I simply drove to the crossing, determined to wait, for several hours, if need be, until a train came along. Figure 4.3 shows the isolated rural crossing at which the collision occurred.

We arrived after sunset. There was no streetlight at the crossing, so it was very dark. Ideally, we hoped to see a train, and more specifically one with black railcars. However, should a train not arrive, the distance at which another target – a pedestrian in dark non-reflective clothing – could be seen would also be an indicator of the degree of visibility of the black oil tanker cars. Although a pedestrian with a white vest is a much smaller target than a railcar, what matters more is the degree of contrast.

Using low beam headlights, I viewed my colleague, who was wearing a black jacket and jeans, walking on the gravel road toward the track area. He disappeared from my view when he was about 75 m (246 ft) away from me. That is significant because at the speed limit of 80 km/h (50 mph), a motorcycle would be covering 22 m (72 ft) every second. A distance of 75 m (246 ft) would be covered in about 3 seconds, barely enough time and distance to

FIGURE 4.3
Daylight photo of crossing at which the collision occurred.

stop. I knew exactly where my colleague was, and so I could see him until he was just barely visible. A driver who did not have that advantage would have much less time. Similarly, it would leave a motorcyclist like Mr Zsoldos, who was not expecting to find a train in the crossing very little time to see a train without reflectors or lighting.

Given that the train was very dark, it would not be very visible. The train might have been made visible by means of the silhouette effect, whereby it was seen as a dark object against a brighter background of the sky. However, the trees beyond the tracks were dense enough in the area immediately behind the crossing that they did not allow a strong silhouette to be created between the train and the sky.

There are so few trains on simple crossings such as these that I knew that we would likely have only one opportunity to see one. And it would likely be moving fast. With only one observation, it would only be possible to gain an impression of how visible it was. I considered several things in choosing how far away we would be from the track to make our one observation. These included: Mr Zsoldos' likely speed, the fact that Mr Zsoldos did see the train but only in time to lay his bike on its side, and the fact that Mr Zsoldos was not expecting the train. For reasons to be discussed in The Science below, I decided to set up our position for viewing at 30 m (98 ft) from the tracks, about 2 seconds' travel time at 50 km/h (31 mph). Although our observation was subjective, it was important to make this observation because it gave me a general impression of the environment. In so doing, it also gave me credibility in the trial that came later because I had experienced to some degree the situation presented to Mr Zsoldos and other motorists who approached the crossing under similar conditions.

Once we were set up in our viewing position, we sat back and waited. Shortly after starting our vigil, Mr Zsoldos, who had been told of our site visit by Mr Mays, drove over in his specially equipped van to visit us. One thing he was able to tell us was that he probably had his low beam headlights on because of the light patches of fog that were present at the time of the collision. After his departure, my colleague and I continued to wait for a train. Since the traffic volume was literally zero and it was a warm summer night, I shut off the car engine, and we waited in the silence of the country, listening for cars and for trains.

An hour passed as we sat quietly lost in our own thoughts. Then suddenly, out of the dark, we heard a voice shout "Jeeezus Christ"! A bicyclist coming up on our rear on this quiet country road saw us at the last second and nearly hit us! The incident demonstrated clearly how dark it was and how difficult it would be to see black oil tankers occupying the railway crossing, considering they were not equipped with lights or reflectors. We continued to wait but in a somewhat more alerted state!

11:00 p.m. was approaching. I turned to my colleague and said I am willing to wait until 11:00 p.m. – if no train arrives by then, I'll have to figure out some other way of determining railcar visibility, perhaps at another similar crossing nearer our office, some 1½ hours' drive away. But luck was with us! At 11:00 p.m. precisely, we heard the sound of a train whistle! I turned the low beam headlights on.

Shortly after the whistle, we saw a short train with truck containers and large gaps between railcars. The railcars on this train, unlike the one that Mr Zsoldos encountered, had reflectors, which made them far more conspicuous than the black oil tanker railcars would have been.

The Science

In this "Science" section I am going to focus on two human factors topics. The first is how far away a range of drivers would have been able to see the black tanker cars. The second is the limitations of a warning sign, without bells, flashers, or gates, to let drivers know they might encounter a train on the track.

Train Visibility

To understand why the crash happened, it is critical to determine seeing distance to the train. In this case, seeing distance was how far away a majority of reasonably alert, unimpaired drivers could see the black, oil tanker railcars, that were passing through the crossing when Mr Zsoldos struck the train. This needs to take into consideration that there were almost no reflectors, there was no lighting except that from the motorcycle's low beam headlight, and there were no active warning devices to draw a driver's attention to the train.

A "majority of drivers" typically refers to 85% or more of reasonably alert, unimpaired drivers, sometimes called the "design driver". Just as one needs to design reach distances to the car dashboard for short drivers, and door frame heights for tall people, one has to design railroad crossings for drivers with short seeing distances. They are members of a "normal" population.

Seeing distance is dependent on the contrast of the object against the background. Black oil tanker cars seen against the gravel road, forest, and overcast sky, provide little contrast and therefore, poor visibility. Seeing distance is also dependent on that individual's sensitivity to contrast which varies considerably.

The distance at which a range of drivers could clearly see the train needed then to be compared with:

i. how much distance Mr Zsoldos would cover in the time (about 1–2 seconds) between perceiving the train was there and initiating braking

ii. what distance would then be required to brake the motorcycle to a stop

As is almost always the case, besides retaining me, a human factors expert, Mr Zsoldos' lawyer had retained a reconstruction engineer. In this case, his or her job would be to determine how much additional distance a motorcyclist in Mr Zsoldos' position needed in order to brake his motorcycle to a stop. This would depend on a number of factors: motorcycle speed, type of braking, nature of the roadway surface (gravel), and position of the motorcycle (upright with tires rolling or sliding on its side) among other things.

There are several ways of coming up with an estimate of seeing distances. One method is to make observations at the site of the object of interest, in our case, black oil tanker cars. Ideally measurements would all be taken with more than one person. This is because people vary in their sensitivity to contrast and therefore, they differ on the distance at which they will see something that is dark and contrasted with something else that is only slightly darker or lighter than the target. For logistical and safety reasons, it was not possible to collect observations at the actual site.

The second method is to make observations of dark-coloured railcars at another railway crossing similar to the one of interest (i.e. no streetlighting, no activated flashers and only warning signs). You would want to measure seeing distance to black railroad cars for a number of people. I had made a set of observations for another case at another railroad crossing of railcars with very small reflectors on them. These observations suggest that a driver using low beam headlights would have a seeing distance of roughly 60 to 100 m (197–328 ft) with an average of 80 m (262 ft). Seeing distance when a driver is not expecting a target has been shown to be approximately half of that measured in an experiment. So, the expected seeing distance for an unsuspecting driver would average about 40 m (131 ft). Research support for this "expectancy correction" is given in this book in Chapter 3 "Pedestrian Lying in Road".

The third method is to depend on research studies which have examined visibility to similar low contrast objects. Not surprisingly, we were unable to find any studies of seeing distance with low beam headlights to oil tanker railcars at night. We did find good studies of seeing distance of pedestrians in dark clothing, seen only by means of low beam headlights. These studies involved 23 young driver subjects and indicate roughly how much seeing distance there might have been for the majority of reasonably alert, unimpaired drivers. They show that a driver using low beam headlights and **not expecting** a target would have a seeing distance of about 10–40 m (33–131 ft) with a median of about 25 m (82 ft). Fifty percent of values lie below the median of 25 m (82 ft), and 50% lie above the median. As for the 10–40 m (33–131 ft) range, 10 m (33 ft) represents the 15th percentile, that is only 15% of drivers have a seeing distance that is worse than 10 m (33 ft), and 85% have a seeing distance that is better than 10 m (33 ft).

Altogether the three methods give us an estimate of about 10 to 40 m (33–131 ft) seeing distance, when considering a wide range of drivers, Mr Zsoldos began to respond to the black tanker cars well within the expected range (see Olson et al., 2010 for a thorough discussion of seeing and expectancy).

Unfortunately, the reconstruction engineer who investigated this crash was unable to determine speed with any accuracy. Therefore, we considered the fact that the road was posted at 80 km/h (50 mph), with an advisory speed sign calling for 20 km/h (13 mph). We also relied on research finding that at another similar crossing, with the same speed limit, but with a STOP sign, the average approach speed was about 50 km/h (32 mph). Consequently, we assumed Mr Zsoldos' speed on the approach was at least 50 km/h (32 mph). At that speed, the motorcycle would be travelling about 15 m (49 ft) per second. The marks on the roadway and on the motorbike, that is the physical evidence, indicated that Mr Zsoldos put the bike on its side in an unsuccessful attempt to avoid the train. Thus, he saw the train, that is he completed his perception-reaction time, activating the brake, during which he would have travelled for approximately 1–2 seconds or about 15–30 m (49–98 ft) before any slowing began. This is why I decided to look at the visibility of the railcars at 30 m (98 ft) at the site visit. It represents the worst case in terms of the greatest distance required to stop. Had the train been clearly visible at 30 m (98 ft), Mr. Zsoldos would have been able to make an emergency stop.

Limitations of Warning Signs for Trains

In this portion of The Science, I will focus on the limitations of "passively protected" crossings which have no flashers, bells, or gates, only warning signs, leaving the course of action up to the driver. Such crossings only have railway crossing "X" signs (see Figure 4.1) to remind drivers to look both ways before crossing, but do not give any real-time information about whether a train is coming, or in the crossing for that matter. Such crossings often also have a suggested low speed posted, as this one did, of 20 km/h (13 mph) or, as was the case at a nearby crossing, a STOP sign. Not surprisingly, these low-speed limits and STOP signs are not very effective at influencing how fast people drive.

Several years after this collision, my colleague Thomas Smahel and I did a study of how fast drivers approached a simple "passively protected" railway crossing. We found that the majority of drivers crossed at speeds higher than those that would allow them to see a dark railcar without reflectors in time to stop. In other words, in the same circumstances as Mr Zsoldos found himself, the majority of drivers would hit the train.

Road and rail authorities recognize that to have a high-speed railway track across a high-speed road is risky. However, there are literally thousands of crossbucks-only crossings in the United States and in Canada. It would be prohibitively costly to put flashers, let alone bells and gates, at every crossing. So, the risk is handled by increasing the amount of protection, that is flashers, bells, and gates, as the number of vehicles and the number of trains crossing in a day increases, and the probability of a train and a road vehicle arriving at the same time, and a collision occurring, becomes more likely.

The magic number for deciding on whether flashers and gates need to be considered is 1,000. When the number of trains times the number of vehicles per day reaches 1,000 then authorities start to consider real-time or activated warnings.

What is the risk of waiting to install flashers until the number of trains times the number of vehicles reaches 1,000? Consider the risk of a crash at a very low railway volume crossing with 10 trains a day. Consider a low-volume road with 100 vehicles a day crossing that rail crossing. At these low volumes the probability of a crash is very low simply because trains and cars are unlikely to be occupying the crossing at the same time.

With 10 trains and 100 vehicles per day and making reasonable assumptions about how long the train would be on the crossing and how long the vehicle would be on the crossing, 10 trains would occupy the track for 20 minutes in a 24-hour period. One hundred vehicles would occupy the track area for approximately 3 seconds each or 5 minutes in total in a 24-hour period. The risk of the two coinciding and a crash occurring is very small.

Unfortunately, such a system does very little for the driver who happens to be at the wrong place at the wrong time.

Conclusions

The central concern of the court case was, "Did a train occupying this crossing pose a hazard to a driver with average capabilities." Mr Zsoldos approached this crossing at night, **with low expectation** of seeing a train, based on his experience of there seldom being a train, especially at night. He had on his low beam headlight rather than the high beam because of patches of fog. There was no lighting besides the headlights. A driver with average seeing distance with low beam headlights (median of 25 m (90 ft) travelling at 50 km/h (31 mph)), would have enough seeing distance to allow him or her to see the dark railcars, but not enough seeing distance to allow him or her to stop before hitting the railcars. Thus, a train occupying this crossing did pose a hazard to a driver with average capabilities.

Countermeasures

What should be done to reduce the risk, without it being too expensive?

A U.S. Department of Transportation study showed that about one quarter of rail crossing crashes involved a vehicle running into the train. The majority of these occur at night, despite the generally lower traffic volumes at night. This strongly suggests that visibility of railcars at night is a contributing factor to crashes. Despite their low traffic and train volumes, almost half of these types of collisions occur at crossings with the simple rail crossing sign and no flashers or gates.

In a laboratory which simulated visibility conditions at night with headlights, U.S. researchers found that railcars with reflectors could be seen 10–17 times further than an unmarked railcar. Adding reflectors to railcars is a lot less expensive than installing flashers, bells, and gates at each crossing (Carroll et al., 1999).

Mr Zsoldos saw the train in time to start stopping but not far enough away to avoid a serious collision. Based on the laboratory studies, adding reflectors to the railcars would have made a big difference to the visibility of the train. Reflectors on railcars had been required in Canada since before the time of this crash (1994). However, most railcars originate from the United States. The United States did not require reflectors on railcars until 2005.

Transport Canada has passed a regulation to require the crossbucks (Railway Crossing) sign and pole to use sheeting that is reflectorized **both front and back**. This means that, at night, drivers approaching from either direction will see the reflective material on the pole and the distinctive X shape of the crossbucks sign.

How does this help the driver? It makes the railway crossings more noticeable, especially at night. When there is a train traveling across the road in the crossing it blocks the driver's view of the back of the crossbucks sign on the other of the tracks intermittently.

First the railcar blocks the view and then the gaps between the moving railcars allow the drivers brief opportunities to see the distinctive shape of the pole and X shape of the crossbuck sign. This produces a flashing effect as the railcars and gaps between go by (see Figure 4.4).

It is not realistic to install flashers, bells, and gates at the thousands of unprotected, mostly rural crossings in Canada and the United States. Depending on a number of factors, gates and flashers cost upwards of half a million dollars per site. However, adding a reflective back on signs is a very inexpensive countermeasure, on the order of $500 per site. Adding reflectors to railcars is also an inexpensive countermeasure which could have assisted Mr Zsoldos.

The Courtroom

Mr Zsoldos sued the railway for its negligence in not inspecting the crossing at night to ensure visibility of trains. Such an inspection, according to the

FIGURE 4.4
Inexpensive countermeasure to attract driver attention to the crossing at night.

judge, would likely have made it evident that more warnings were necessary and could have prevented this crash. The rail company was found 75% liable, and Mr Zsoldos, because he was judged to have been drinking and driving too quickly, 25% liable. The railway company appealed the judge's decision. They lost for a second time.

References

Carroll, A. A., Multer, J., Williams, D., and Yaffee, M. A. (1999). *Freight car reflectorization* (Report no. DOT/FRA/ORD-98/11). U.S. Department of Transportation.

Gabree, S. H., Calabrese, C., Young, J., Mejia, B., Prescott, J., & John, A. (2019). *Review of highway-rail grade crossing incidents 1986–2015: Trends and potential safety gaps* (Report No. DOT/FRA/ORD-19/30). Federal Railroad Administration, Office of Railroad Policy and Development.

National Highway Traffic Safety Administration (NHTSA). (2003). *Test track examination of drivers' collision avoidance behavior using conventional and antilock brakes* (Report No. DOT HS 809 561). U.S. Department of Transportation.

Olson P., Farber E., & Dewar, R. (2010). The visibility provided by vehicle lighting systems. In *Forensic aspects of driver perception and response* (3rd ed.). Lawyers & Judges Publishing Company.

5

Eyewitness Evidence in a Rear-End Crash

Introduction

Most lawsuits concerning motor vehicle collisions depend in part on eyewitness evidence to determine the cause. How reliable is such evidence? In this chapter, we consider a rear-end crash in which two vehicles were following one another when the first one slowed to make a left turn into a commercial driveway. The left-turning driver claimed he was stopped, for several seconds waiting to turn when he was suddenly struck from behind by another vehicle. The following driver said, on the contrary, the driver he was following slammed on the brakes at the last moment and he could not avoid a collision.

How can two people who witness an event come up with such different stories? It has to do with the first issue I consider in The Science section below, which is our ability to take in or "process" information and remember it accurately. I also consider the many ways in which witnesses' memories can be biased.

The second Science issue I consider is the time and distance that need to be maintained between two vehicles, one behind the other, to ensure that the following driver can stop in time should the driver in front have to stop quickly. This depends in part on the following driver's perception-reaction time. It also depends, among other things, on the state of the brakes and the friction with the road surface.

Collision Description

According to the police report, the collision took place in a suburban area in Red Deer, Alberta, in the summer at about 5:00 p.m. The driver who was rear-ended was Mr John Turner, aged 70. He was driving a Volvo eastbound on Smith Road. The second driver, Mr Scott Gretski, aged 26, was driving a Ford F-150 eastbound on Smith Road, following the Volvo. Mr Turner stopped to

DOI: 10.1201/9781003564805-5

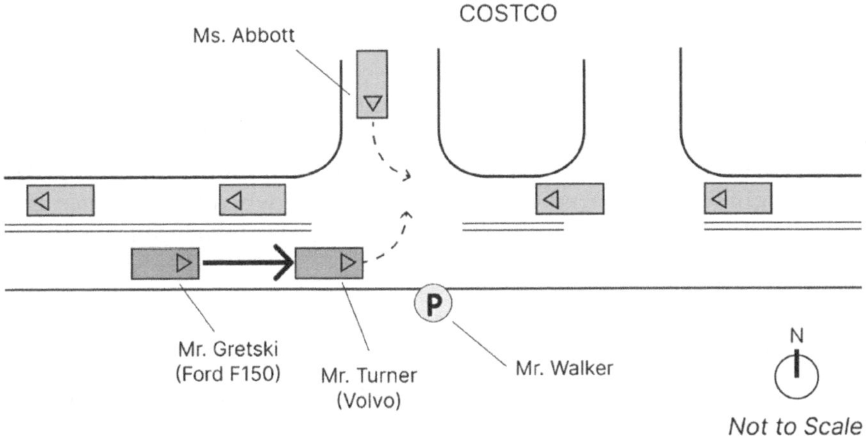

FIGURE 5.1
Positions of the involved parties just prior to collision.

turn left into the driveway of a Costco store and his car was rear-ended by Mr Gretski's Ford F-150.

Ms Abbott, aged 42, was an independent witness waiting to turn left onto Smith Road. Mr Walker, aged 15 years, was an independent witness on foot waiting to cross Smith Road. Figure 5.1 shows the positions of the involved parties. The collision occurred under an overcast sky, in clear and dry weather conditions. Smith Road has a posted speed limit of 50 km/h (30 mph).

Evidence Given By Involved Parties

Typically, evidence given by the involved parties is recorded in several forms. Most often, drivers make a statement to the police shortly after the collision. The driver may also give a statement to his or her insurer. If the case does not settle, an examination for discovery (called a deposition in the United States) is conducted. This allows the opposing lawyers to find out what the various witnesses are likely to say in court, in hope of reaching a settlement before the prolonged and expensive process of a trial. This usually occurs 1 or more years after the collision. Finally, there is the evidence given at trial, also 1 or more years after the collision. This can be a long and difficult process for the parties involved in the crash, especially those who are likely to be found at fault. I have had cases in which more than 10 years elapsed between the crash and the settlement. If there are appeals, the process can take even longer. Criminal cases (e.g. driving under the influence, dangerous driving causing death, and so on) are dealt with much more expeditiously than civil cases. The latter involve deciding what percentage driver or other road user is at fault in determining which insurance company will pay how much.

Sometimes, and especially so with examinations for discovery, there is a long gap between the time of the collision and the recording of witness evidence. This is not good when one is depending on accurate memories. Also, more weight is often placed on the latest evidence when the evidence given immediately after the event is more likely to be closer to an accurate description of what happened.

Mr Turner's Evidence

In his statement to the police, the driver of the Volvo which was rear-ended said that, at the time of the collision, he had been driving on Smith Road with the intention of heading towards the Costco store in Smith Plaza. He was with his 14-year-old grandson who was showing him the way as he had never been to the Plaza before. He recalled the oncoming traffic on Smith Road to be busy. Upon approaching the driveway to Costco, "When my grandson pointed out the driveway I slowed and put my turn signal on and then stopped." In his Examination for Discovery, with respect to the amount of time that Mr Turner had been stopped before he felt the impact with the following vehicle, he indicated that he was not certain how long it had been. He believed that he was stopped with his left signal on, waiting to turn left into the driveway, while at least four vehicles passed in the opposite direction. He believed that the fourth vehicle had passed, or was about to pass, when his vehicle was struck from behind. He was completely unaware of the following driver until he was struck.

Mr Gretski's Evidence

The driver who rear-ended the Turner vehicle had quite a different story. Because of his injuries (he was not wearing a seatbelt), Mr Gretski was not able to give a statement to the police until 2 months **after** the collision.

He said that he had turned onto Smith Road at Main Street and had quickly caught up to the Volvo (driven by Mr Turner). **He followed the Volvo for 1 or 2 minutes before the collision occurred**. At the time of the collision, he was travelling at about **30 km/h** (19 mph) on Smith Road, behind the Volvo. He said, "The Volvo was travelling at the same speed as I was, and I was **about 3 to 4 car lengths behind him**." He said that "Suddenly, without any signal or warning, the Volvo driver jammed on the brakes, standing the rear of the Volvo up." He was not able to stop in time to avoid a serious collision.

According to his Examination for Discovery, given 3 years after the collision, Mr Gretski said that, at the time leading up to the collision, he had just left work and was on his way to pick up his friend to play hockey. He was a little late.

Mr Gretski corrected earlier information he had given the police that he had followed the Volvo for **1 or 2 minutes before the collision occurred**. He said he was no longer certain how long he had followed the Volvo, but he thought it was no more than 30 seconds. This correction may have been made because he timed the route himself and realized that he would not have taken **1 or 2** minutes to travel the distance in question. If people involved in crashes do not do this themselves, they can be sure that any expert involved in the case will have checked the length of time it takes to get from point A to point B.

Mr Gretski said when he first noticed the Volvo in front of him, and started following it, he was **about two or three car lengths from it. When he first noticed the brake light on the Volvo in front go on, he was 30 ft behind it.**

Memory fades with time. Note the long time periods that had elapsed between the event of the collision and the request for details of how it happened (2 months until the report to the insurer and 3 years until the Examination for Discovery). Also note the different estimates for how far ahead the Turner Volvo was in front of the Gretski vehicle – **three to four car lengths (Statement to Police), two to three car lengths, 30 ft (Examination for Discovery)**. This illustrates the uncertainty of witness evidence about distances (and times). If only one estimate in one set of units (e.g. feet, metres, car lengths, lengths of a football field (a favourite measure of some male lawyers)) is requested, one can inappropriately depend on it being more accurate than it really is. By inappropriately I mean judging liability based on it. However, if no other evidence is available, liability can be determined based on single, inaccurate estimate of distance (or time).

An issue in Canada is the confabulation of metric (e.g. km/h) and U.K. (and U.S.) (e.g. mph) units. Sometimes it is difficult to tell whether a witness meant 50 km/h or 50 mph A related issue here is the definition of "car length". Some may think of it as literally the length of a car, which of course varies. Others may think of it as the car length plus the space between vehicles. At any rate, eyewitness estimates are to be treated with caution. Physical evidence such as video recordings from dashcams or security cameras with a view of the collision scene, which are more common recently, are greatly preferred.

The actual distance between the leading Volvo and the following Ford F-150 just prior to the collision would be critical in the defence of the Ford F-150 driver. If it was too short his driving could be seen to be "aggressive" and a contributing factor to the rear-end crash. If it was much longer, the question could be raised as to why, with all that time, he was not able to stop in time.

In my experience, there is a very human tendency for witnesses to make estimates in their favour, and to claim, for example, that they were going the speed limit. It is best for involved parties to avoid making estimates unless you had happened to look at the speedometer at that exact moment you are asked about. If your estimate does not agree with the physical evidence your attempt to estimate in your favour can backfire. Furthermore, an involved party can, without realizing it, give evidence that though it appears to be more favourable to themselves, it actually provides evidence that leads to them being found at fault.

Mr Gretski said that there were two driveway entrances to the Costco Store. **The Volvo driver tapped the brake as he got closer to the first entrance.** Mr Gretski thought at first that he (Turner) was intending to turn at the first entrance. However, as he (Turner) got closer to the first driveway, he did not seem to be slowing enough. Mr Gretski then assumed that the Volvo driver was going for the second entrance. Suddenly the Volvo driver slammed on the brakes, stopping in the middle of the first entrance driveway after all. Mr Gretski did not see a turn signal on the Volvo at any time.

Mr Gretski said that he had been travelling at no more than 35 km/h and that his foot was hovering over the brake pedal after seeing the Volvo brake lights for the first time.

Ms Abbott: Independent Witness Driver

In her statement to the police on the day after the collision, Ms Abbott said that when the collision happened, she was exiting waiting to turn left onto Smith Road. "I was watching to the right as there was more traffic coming from that direction. There was a gap in the traffic and I was about to pull out when a white Volvo going eastbound suddenly slammed on the brakes and was trying to turn left into the driveway of the Costco store." **She noted that the driver of the Volvo did not signal.** As for the pick-up truck following the Volvo, the Ford F-150 tried to swerve to miss the Volvo but hit the back of it. Ms Abbott stated that the car following the Volvo was travelling "no more than about 35 to 40 km/h".

Mr Walker: Independent Pedestrian Witness

In a statement given 2 years after the collision Ryan Walker said that on the day of the collision, he was walking through the parking lot across Smith Road. He said he was about to run in front of the Volvo (driven by Mr Turner), but he saw the Volvo slowing. Contrary to Mr Gretski's evidence Mr Walker said he saw the Volvo's turn signal flashing, and so he decided not to run across Smith Road in front of it.

He said the Volvo was stopped and another car (the Ford F-150, driven by Mr Gretski) was coming up behind it. The Ford F-150 slammed into the back of the Volvo. He estimated that 4 or 5 seconds went by between his deciding not to run and the Volvo slowing down or stopping before the collision. This implies that Mr Gretski would have had visual evidence for at least 4–5 seconds that Mr Turner was going to slow down sufficiently to turn or stop. Mr Walker said that he had a clear and unobstructed view of the entire collision.

Site Visit

On a clear summer afternoon, I visited the site and observed the area of the crash from the perspective of the driver witness (Ms Abbott) and the pedestrian witness (Mr Walker). In order to assess how good a view each witness had, I measured how far away they were from the collision and what their view was. Ms Abbott would have been about 50 m (70 ft) north from where the collision occurred. She said there was a gap in the westbound traffic and she was about to turn when the crash unfolded in front of her.

I then looked at the distance from the site of the collision, generally called "the area of impact", to where Mr Walker was walking. He would have been approximately 10 m (16 ft) from the area of impact when he stopped and turned to watch the approaching eastbound traffic. He would have had a good view of the collision.

The Science

In this "Science" section I am going to focus on two human factors issues.

The first issue is our ability to process information and remember it. When we watch an event like a collision unfold, there are many aspects we could **notice** out of the many that we **see**. I have bolded the words "**notice**" and "**see**" to emphasize the difference between what is available to be **seen** and what is **noticed**, the latter being much more limited than the former (see discussion in Case 1). Also, people **notice** different things. For example, during an intersection collision, one witness may **notice** that the traffic light had just turned green when the driver entered the intersection. Another witness to the same collision may **notice** that the pedestrian who was struck was using his cell phone, but may have no idea of the state of the traffic signal at the time.

I will also describe some of the aspects of selecting and storing information that affect how accurate and how detailed a memory is.

The second human factors issue concerns the distance you need to keep between your vehicle and the vehicle ahead that would guarantee that you would be able to stop in time should the driver in front of you stop suddenly.

Information Processing, Memory, and Biases

Human attention and abilities in information processing are limited. In this case, "information" refers to details of the road scene and road user actions, for example: "the black van went through a red light" or "the driver looked left just before crossing the intersection." There are only so many observations that can be made in the few seconds during which a collision develops and occurs.

On the plus side, attention, like a searchlight, can be switched rapidly from one information source to another. However, humans only attend well to one information source at a time. For example, one witness might pay attention to the actions of one driver, and another to a different driver, and still another to the actions of the pedestrian or to the state of the traffic lights when the vehicle entered the intersection.

In the moments leading up to a collision there is a great deal of information to be **noticed** by eyewitnesses out of what can potentially be **seen**. This includes the paths of the vehicles, state of headlights, etc.

Information can be measured in terms of the number of yes/no questions that can be answered by a witness who is exposed to a single scene. For example, the yes or no answer to the question "did the driver go through a red light" provides one bit of information. The more yes/no questions that can be answered, the more information is being transmitted and potentially retained in memory.

Humans can only extract a small proportion of the available information from a road scene. It has been estimated that in any one second, over one billion bits of information are directed at the sensory system (the eyes, the ears, the nose and so on) (Grandjean, 1988). This means it would take on the order of a billion yes/no questions to fully describe 1 second's worth of a road scene. With respect to driving, most sensory information is visual. Out of the enormous amount of information that can potentially be **seen**, only about 16 bits/sec of new information is consciously **noticed**.

Information must not only be **noticed** but also stored in memory. Many factors affect this. One of these is the amount of time that passes between an event and a witness being asked to recollect the event. Another is observer bias due to whose actions are being reported – your own or those of someone else. Yet another is the observational point of view. What you can **see** and report from one position relative to a collision may be different from what you can **see** and accurately report from a different position.

Given the limitations in driver information processing, as well as biases, it is not surprising that witnesses to a collision frequently disagree on details of the collision or make statements that are inconsistent with physical evidence (Loftus, 1996). For example, a witness may be influenced to make an estimate of the speed of a vehicle, which is influenced by a tragic outcome to the driver actions, that is much higher than the speed reported by an electronic data recorder (the physical evidence) in that vehicle.

Constructing Memory

According to psychologists, memory of an event, especially one that is fleeting, is seldom an exact reproduction of the event. Memory is more accurately described as a reconstruction of events from a series of memory fragments. A series of brief images are strung together to create the memory of an event or conversation. In recalling an event, we retrieve the fragments and reconstruct the event from the series of fragments. A good example of this is when we try to remember a conversation. "We do not have word-for-word copies of the conversation stored in our memory. What we have are general facts or ideas about what was said" (Bourne & Ekstrand, 1979). We fill in the missing parts of the memory of the event with likely connections, believing that the missing parts that were filled in are as accurate as the parts that were correctly remembered.

Expectancy Bias

An example of the expectancy bias that can occur while memory is being constructed is shown by the following incident. A vehicle is observed as it stops to make a left turn, and then it turns. The brake lights must have been on, so that is "remembered" by a witness because it is expected. But the reality could be that the brake lights were not on. This may have been because they were not working, and this could be proven with physical evidence. This is an example of a bias due to expectation. This bias can lead to inaccuracy which can occur when people try to remember and make sense of the details of a traffic collision by filling in what is expected.

Delay in Recollection

Typically, police statements are taken shortly after a collision: usually, a matter of hours or days afterwards. Examinations for discovery, on the other hand, are more typically taken several years after a collision and are generally much more detailed.

Research shows that memory fades and is easily corrupted over time. Thus, the ideal time to collect information about a crash is immediately afterwards and not several years later. The more immediate the collection of information, the more likely it is to be accurate. In part this is because participants and witnesses have not had time to rationalize anybody's actions and are simply reporting without much bias. Nor have witnesses had the opportunity to discuss what happened with others and be influenced by hearing incorrect versions of what happened. Witnesses to a collision may be asked several years after an incident about details that were unimportant at the time. An example is the question "which way did the driver look when turning just prior to impact, left or right?" The witness may answer the question, and the judge and jury may rely on the answer. However, it is hard to believe that such a detail, which was unremarkable at the time, before the collision occurred, would be remembered accurately several years after a crash.

Hindsight Bias

A witness who knows that the outcome of a sequence of events was a crash is more likely to consider it to have been predictable than it would have been for the involved driver witness who did not know the outcome. This is often summed up in the phrase "hindsight is 20/20" (Dilich et al., 2006).

Thus, there are inaccuracies which result from our limitations in collecting information. There are inaccuracies arising from the way we construct memory. And there are numerous types of observer bias, such as expectancy and hindsight bias as discussed above.

Perception-Reaction Time

The second science issue I want to discuss is what distance or travel time you need to keep between your vehicle and the vehicle ahead that would guarantee being able to stop in time should the driver in front of you stop suddenly. This distance/travel time depends on the lag between you perceiving that the vehicle ahead is braking and your application of the brakes. The more slowly you respond, the longer you will travel without braking, catching up to the vehicle ahead. Perception-reaction time refers to the time it takes from seeing a hazard developing to putting your foot on the brake. Time/distance to come to a stop is not included in perception-reaction time. An experimental study based on almost 200 drivers, stopping in response to a vehicle suddenly pulling across their path, estimates that the majority of drivers (85%) can initiate braking within about 1.5 seconds. Experimental studies also show that having your foot hovering over the brakes reduces perception-reaction time to about 1 second for the large majority of drivers (Mazzae, Barickman, Baldwin, & Forkenbrock, 1999).

Analysis

Three drivers and one pedestrian were involved in or witnessed this collision. One driver and one pedestrian were independent witnesses (Ms Abbott and Mr Walker); two drivers were involved in the collision (Mr Turner and Mr Gretski). Figure 5.1 earlier in this chapter shows the positions of these people relative to each other.

Memory is affected by exposure time. In this case, exposure time was brief: the events in this collision occurred over a few seconds. Not surprisingly, given the number of observations that could be made, the short time period and the unexpectedness of the event, the witnesses disagree about the details of what happened.

The most important disagreements among the witnesses to our collision concern:

- whether or not Mr Turner's turn signal was on
- for how long Mr Turner was stopped, if at all, before his Volvo was struck
- Mr Turner's braking pattern

Let us consider each of these in turn.

1. Was the Turn Signal On?

There is some disagreement as to whether Mr Turner had his turn signal on. Mr Gretski, the driver who struck the Turner vehicle, as well as independent driver witness, Ms Abbott, say that it was not on. Mr Turner, and the pedestrian witness, Mr Walker, say that it was on.

First with respect to the passage of time, Ms Abbott's evidence was given much closer in time to the actual event: the day after the collision, while the pedestrian witness' statement was given 2 years later. The longer the time after an event occurs that a memory is elicited, the less likely it is to be accurate.

Second, with respect to the observer's point of view, the independent driver witness, Ms Abbott, was at a greater distance (50 m) (80 ft) from the involved vehicles than was the pedestrian witness, Mr Walker. Ms Abbott's view of the eastbound Volvo may have been blocked to some degree by one of the westbound vehicles approaching the collision area where Mr Turner was waiting to turn from eastbound to southbound. By contrast, Mr Walker was standing 8 m (26 ft) from the area of impact with an unobstructed view. Thus, while Mr Walker had the better view, Ms Abbott's evidence was given much closer to the time of the collision. The actions of both witnesses were affected by Mr Turner's intended turn, so they would have had good reason to **notice** and remember the turn signal.

Based on the evidence given by four people, there is no compelling reason to prefer one set of observations (two said signal off) to the other (two said signal on).

2. What Was the Braking Pattern of Mr Turner's Volvo?

Ms Abbott had a good view of the approaching Volvo. At the time of the crash, she was waiting to turn left onto Smith Road. She said she was watching to the right as there was more traffic coming from that direction. She watched the Volvo as it approached the driveway to Costco. According to her, the driver of the Volvo suddenly slammed on the brakes and was trying to turn left into the driveway to Costco. Mr Gretski, the Ford F-150 driver, tried to swerve but rear-ended the Volvo.

Mr Walker said that he had been walking on the sidewalk intending to cross the road in front of the Volvo. He was about to run across Smith Road in front of the Volvo but he saw it slowing and so he decided not to.

He said that just before the crash, the Volvo had stopped and another car was coming up behind it quickly. It slammed into the back of the Volvo. He estimated that 4 or 5 seconds passed between his deciding not to run and the Volvo slowing before the collision. This is an example of the common phenomenon of eyewitnesses giving slightly different versions of what happened. First Mr Walker says he was about to run across Smith Road in front of the Volvo but he saw it slowing and so he decided not to. Then he says his decision not to run was made 4 or 5 seconds before he saw the Volvo slow. He says he had a clear and unobstructed view of the entire collision.

The driver of the Volvo believes that his Volvo was stopped with its left signal on, waiting to turn left into Costco, while about four vehicles passed in the opposite direction. He believes that the fourth vehicle had passed, or was about to pass, when his vehicle was struck from behind. Assuming vehicles passed every 2 seconds, the recommended spacing in the Alberta driver's handbook. this would occupy approximately 8 seconds. In other words, he would have been stopped for about 8 seconds before his car was struck.

(In retrospect this is not a good method of coming up with likely time-headways. For one thing it is biased against the other side. This is because drivers typically do not obey the 2-second guideline but choose shorter time headways. It would have been preferable to videotape platoons of traffic arriving from the upstream traffic light at the site and then measure the average time gap. The time it took to wait for four cars to pass depends on traffic volume and could easily have been 4 seconds rather than 8 seconds. The less time Mr Turner waited, the more likely he would have been seen to have stopped abruptly.)

There were two driveway entrances to the Costco store. Mr Gretski said that the driver of the Volvo in front of him tapped the brake as he got closer to the first entrance, and he initially thought that he was intending to turn at the first entrance. However, he did not seem to be slowing enough so he assumed he was going for the second entrance. But suddenly he slammed on the brakes, stopping at the mid-point of the first entrance driveway after all. He did not see a turn signal at any time.

Thus, one independent witness, the pedestrian, supports Mr Turner's description of the braking event. The other independent witness, Ms Abbott, supports Mr Gretski's description.

3. How Long Was the Volvo Stopped Before It Was Struck?

In his Examination for Discovery, 3 years after the collision, the Volvo driver initially said he was stopped for 1 minute or so waiting for the oncoming traffic to pass. When questioned further about how he determined this, Mr Turner said he was not really certain how long he had been stopped. The traffic was busy and he estimated that he was stopped with his left signal activated, waiting to turn left into the Costco store, for the duration of several vehicles – maybe four, passing his stationary vehicle in the opposite direction.

In his statement made 2 years after the collision, Mr Walker (the pedestrian) said that 4 or 5 seconds went by between his deciding not to run and the Volvo getting hit. He did not indicate if or for how long the Volvo was stopped. It could have been momentarily. Ms Abbott reported that the Volvo driver slammed on the brakes, suggesting that Mr Turner was not stopped or had just stopped when hit.

In a lawsuit involving two drivers like this one, the actions of both drivers, whether they are at fault or not, are explored to see how each could have contributed to avoiding the crash.

4. Actions of the Mr Turner, the Volvo Driver

The versions given by Mr Turner and Mr Gretski of how the crash occurred are very different. Mr Turner claimed that he was stopped during the time it took about four vehicles to pass him. As indicated above, this could occupy approximately 8 seconds. Mr Gretski said that he had been following three to four car lengths behind Mr Turner when suddenly, without any signal or warning, the Volvo stopped violently in front of him.

Both drivers show observer bias. Being stopped, as claimed by Mr Turner, is not supported by Ms Abbott, an independent driver witness who was waiting to turn left and says she watched the Volvo approach. An action of following at three to four car lengths, as claimed by Mr Gretski, is not supported either. He was obviously attentive to the car in front, given his detailed description of what happened (his foot hovering over the brake, an initial touching the brakes and then slamming on the brakes). If he was really travelling at three to four car lengths behind the Volvo, he should have been able to stop in time, as we shall see.

5. Actions of Mr Gretski, the Ford F-150 Driver

Perception-reaction time refers to the time it takes from seeing a hazard developing to putting your foot on the brake. Time to come to a stop once the brakes are first activated is not included in perception-reaction time. A large experimental study estimates that the majority of drivers can initiate braking within about 1.5 seconds. Note that 1.5 seconds is the time for perceiving the hazard and just beginning to respond. It does not include time for the vehicle to decelerate to a stop.

If the Ford F-150 was travelling at least 1.5 seconds behind the Volvo, and if Mr Gretski could decelerate as quickly or faster than Mr Turner, he should have been able to stop without colliding with him and with time/distance to spare.

Even if Mr Turner did not have his signal on, his brake lights would have been on indicating he was slowing. Indeed, in his Examination for Discovery Mr Gretski said that his foot was on the brake "since it seemed like he (Turner) was about to make a turn". In this case experimental studies show that having your foot hovering over the brakes reduces perception-reaction time from 1.5 seconds to about 1 second for the large majority of drivers.

Government authorities (in the form of the Driver's Handbook) caution drivers to maintain a 2-second headway with the vehicle in front. You can measure your time headway by observing when the vehicle ahead of you crosses a marker such as a stop bar, and then counting the number of seconds (by saying "one one thousand, two one thousand", etc.) until your vehicle crosses the same marker. One second is about the length of time it takes to say "one one thousand".

Adopting a 2-second headway would have allowed the majority of drivers about 1 second for perception-reaction time, and 1 second to respond to the Volvo driver's brake lights by braking or swerving. This would leave time and distance to spare. The majority of drivers would be expected to be able to avoid a collision, whether or not the Volvo turn signal was on. Had the Volvo stopped less precipitously, Mr Gretski would have had even more time to brake, thereby avoiding a collision.

Mr Gretski says that he was 30 ft from the Volvo when he touched his brakes for the second time. Thirty feet is equivalent to about 10 m. At 35 km/h (21 mph), the Ford F-150 was travelling about 10 m (16 ft) every second. Since he was unable to stop, Mr Gretski must have been less than 1 second behind the Volvo, not the 2 seconds recommended. With this less-than-1-second headway, he would have barely had enough time to perceive and react to the Volvo, let alone to actually stop the vehicle.

Conclusions

The sequence of events involved in the crash was complex and the events occurred within a few seconds. The witnesses were, of course, not prepared in advance to make specific observations but simply reported what they remembered after the incident. Human limitations in information processing and memory are such that only some of the many details of an event seen during a brief period of time will be remembered, and some of those details will be incorrect. Observers tend to remember fragments of information and to rely on expectations in reconstructing what transpired. In the case of this collision the time elapsed between the collision and the witnesses' statements differed, with Ms Abbott's evidence being given closest to the time of the collision. The vantage points of the independent witnesses were different as well as their observational bias.

Based on the pedestrian witness' better vantage point (an unobstructed view), closer proximity, and the direct effect that the observation of the turn signal had on his actions (deciding to wait to cross the driveway entrance), I was tempted to give more weight to the pedestrian's evidence concerning the turn signal than to the driver witness. However, the pedestrian witness could have seen the Volvo slowing to turn and simply assumed, rather than actually saw, that his turn signal was on. The pedestrian witness did not indicate if or for how long he was stopped. It could have been momentarily.

Ms Abbott reported that the Volvo driver slammed on the brakes, suggesting that Mr Turner was not stopped and waiting to turn when hit.

Both Ms Abbott and Mr Gretski reported passing westbound traffic before the crash, supporting Mr Turner's evidence that he waited for several westbound vehicles to pass.

Whether or not Mr Turner had activated his turn signal, and whether or not he had stopped abruptly, had Mr Gretski been following at the recommended 2-second headway, he would have had 2 seconds for perception-reaction time. The majority of drivers, whose feet are at the ready (i.e. covering the brake), would be expected to need about 1 second in this situation.

Had the Volvo driver stopped less precipitously, the Ford F-150 driver would have had even more than 2 seconds for perception-reaction time. If the Ford F-150 driver had had his foot on the brake before the Volvo driver stopped, as he stated, he should have been able to respond within 1 second and would have had even more time to complete a braking response. In my opinion, both drivers contributed to this collision, Mr Gretski by driving too closely behind Mr Turner, and Mr Turner by stopping too precipitously, possibly without using his turn signal.

Countermeasures

Rear-end crashes are very common. About 25% of intersection crashes are rear-ends. What can be done to reduce the risk?

More and more vehicles are equipped with automatic warning or braking. A forward collision warning system (FCW) does just that – it warns of an impending collision with a vehicle in front. It is left up to the driver to brake. A second more sophisticated system (low-speed automatic emergency braking (AEB)) deals with stop and go situations but only operates at speeds up to 30 km/h (19 mph). It does not warn, it brakes. A third system (FCW with AEB) warns and brakes automatically (Cicchino, 2017).

These features reduced rates of rear-end striking crash involvements with injuries by 20%, 45%, and 56%, respectively, by FCW alone, low-speed AEB, and FCW with AEB.

References

Bourne, L. E., Jr., & Ekstrand, B. R. (1979). Human learning and memory. In *Psychology* (3rd ed.). Holt, Rinehart and Winston.

Cicchino, J. B. (2017). Effectiveness of forward collision warning and autonomous emergency braking systems in reducing front-to-rear crash rates. *Accident Analysis & Prevention, 99*, 142–152.

Dilich, M., Kopernik, D., & Goebelbecker, J. (2006). Hindsight judgment of driver fault in traffic accident analysis. *Transportation Research Record, 1980*, 1–7.

Grandjean, E. (1988). *Fitting the task to the man: A textbook of occupational ergonomics*. Taylor & Francis Ltd.

Loftus, E. F. (1996). *Eyewitness testimony*. Harvard University Press.

Mazzae, E. N., Barickman, F., Baldwin, G. S., & Forkenbrock, G. (1999). Driver crash avoidance behavior with ABS in an intersection incursion scenario on dry versus wet pavement. *SAE Transactions, 108*, 2340–2352.

6

Pedestrian Struck on Traffic Signal Change

Introduction

A teenager, Chelsea Butler, busy with her cell phone, crossed an intersection against a red light. She was struck by a driver, Anaya Abbas, on the cross street, who had entered the intersection just as her red traffic light turned green. The driver's view of the pedestrian was partially blocked by an SUV in the adjacent lane. The sun was rising and the glare of the sun beside the signal was making it (the traffic signal) difficult to see. The human factors issues explored in The Science section of this chapter deal with:

1. detecting a pedestrian crossing on her (the pedestrian's) red light for a driver preoccupied watching for her light to turn green when the sun was beside it creating glare
2. detecting a pedestrian when they are more than 10° off the driver's line of sight
3. view blockage for the driver
4. perception and reaction time of the driver

Also explored in this case is how the human factors expert and the engineer who does the collision reconstruction work together to understand how the collision came about.

The Mandate

My mandate was to explore two questions:

1. Would a normally cautious driver have noticed Chelsea in the few seconds she would have been visible **before** she became obstructed by the Hartung SUV?

DOI: 10.1201/9781003564805-6

2. Would a normally cautious driver have enough time to stop **after** his/her view of Chelsea was unobstructed?

The Collision

One fine spring morning, just after sunrise, Ms Chelsea Butler (aged 14) was on her way to an early morning class at her high school. She was late. As she approached the traffic light at Larch Avenue and Main Street, she was texting a friend she was going to meet on the way to the class. When she got to the intersection, she saw a stopped SUV in the left turn lane and a Jeep in the far lane. She told the police: "It (the SUV) was waiting for me. I jogged across. I thought I could make it easily. Then a red car suddenly shot out between the SUV and the Jeep. It was really fast." Chelsea was seriously injured. The speed limit was 50 km/h (30 mph). See Figure 6.1 showing positions of involved parties and area obstructed by SUV just prior to the collision.

At first Chelsea said she could not remember if she looked at the traffic signal when she entered the crosswalk. Later she said it was only a few seconds after entering the crosswalk that she looked up at the light. Once she saw the amber light she began to walk quickly. She admitted she was carrying a cell

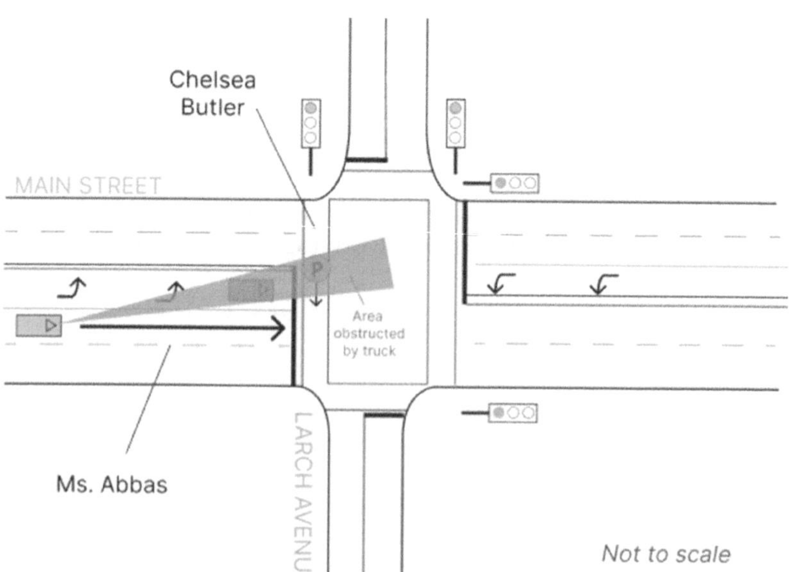

FIGURE 6.1
Positions of involved parties just prior to collision.

phone when she was struck but insisted to the police that she had finished texting before she entered the crosswalk.

While Chelsea was crossing, Ms Anaya Abbas, in a red Toyota, was driving east on Main Street towards the intersection with Larch Avenue, on her way to work at the local hospital. As she approached the intersection, the light at Larch Avenue was red, so she slowed down, not sure whether she would have to stop or not. The light had been red since she passed through the last intersection, so she knew from experience that it was close to changing. As she neared the intersection, her light changed from red to green. Ms Abbas told the police that she was travelling at about 10 km/h (6 mph) in the passing lane (lane two) before the light changed. Then she sped up to perhaps 15 km/h (9 mph) when she entered the intersection. Her view of Chelsea was blocked by the SUV stopped in the turn lane on her left. There was a Jeep on her right just starting to move after the light changed. She was about to pass between the SUV and the Jeep when Chelsea suddenly appeared in front of her. She was not able to hit the brakes before striking her.

Ms Abbas said she was watching the traffic signal carefully because the closeness of the early morning sun to the signal made it difficult for drivers to see when the light changed from red to green. She said that the SUV in the left turn lane to her left blocked her view of Chelsea until the last second. Ms Abbas said although she braked, it was not until after she struck Chelsea that she was able to do so. This information about where she braked is useful to the reconstructionist. The Toyota (Abbas vehicle) was not moved after it initially came to rest. Assuming an emergency stop, the longer the distance between where Chelsea was struck and where the Abbas vehicle came to rest, the higher the speed must have been when Ms Abbas hit the brakes. The driver's evidence is useful to the human factors expert as well as to the reconstructionist. It tells us that the driver may have had time to carry out some part of the perception-reaction process – up to 1 to 1 ½ seconds, but no time for braking before she struck Chelsea.

In her statement at the scene, Ms Lesley Cunningham, a pedestrian, said she had been walking behind Chelsea and got to the corner just after the light had turned amber. She was concerned because she saw Chelsea walking and looking down at her cell phone when she entered the crosswalk at the intersection.

In his statement at the scene, the SUV driver, Mr Hartung, stated that he had stopped at a red light at Larch Avenue waiting to turn left. The light turned green. As he started to go, he was passed by a red Toyota (Abbas vehicle). He heard a "bang" and saw a person "roll into the intersection". He said the pedestrian was crossing within the crosswalk. He was not sure from which direction she came. (The reconstructionist reported that since Chelsea's injuries were on her right side and Ms Abbas was travelling eastbound, Chelsea must have been southbound.) Chelsea was wearing a grey jogging suit.

Information from the Reconstructionist

It is usually the case that the human factors expert must depend on a reconstructionist, who has been hired by the same lawyer, for certain key pieces of information such as the speed of vehicles at impact. It is also sometimes the case that the reconstructionist depends on the human factors expert, particularly in cases where the determination of perception reaction time is complex, or where driver workload, fatigue, distraction, and other more psychological issues come into play.

In the case of this collision, important information provided by the reconstructionist was:

- the estimated speed of the Abbas Toyota on the approach to the collision point
- the estimated speed of Chelsea as she walked or jogged across the intersection
- the amount of time Ms Abbas had an unobstructed view of Chelsea after she passed in front of the Hartung SUV
- the angle Chelsea would have been seen at in Ms Abbas' peripheral vision
- the colour of the traffic signal while Chelsea was crossing

The speeds of the Abbas Toyota and Chelsea were estimated and then used to calculate the unobstructed view time, the angle at which Chelsea would have been seen, and the traffic signal state.

Speed of Abbas Toyota: Unbeknownst to most drivers is that any vehicle with an airbag has an electronic data recorder. This recorder is capable of recording speeds and other measures for the last 5 seconds before a sudden slowdown due to hitting something, causing the airbags to go off. Ms Abbas did have a recorder; however, a pedestrian does not have sufficient mass to trigger the airbags. Consequently, traditional means had to be used to estimate speed of the Toyota at impact. Based on the location and amount of damage to Ms Abbas' vehicle, on the path and distance followed by Chelsea's body after she was struck, on skid marks and on coefficient of friction of the pavement, the reconstructionist estimated Ms Abbas had been travelling at 25 km/h (16 mph) just prior to striking Chelsea. Note that this is half of the speed limit of 50 km/h (31 mph).

Chelsea's Speed: Another important factor is the speed at which Chelsea was walking or jogging. (This may be determined by the human factors expert or by the reconstructionist.) In this case Chelsea's speed was determined by what an independent witness said about her speed as well as what Chelsea herself said.

Ms Cunningham, a pedestrian, said she saw Chelsea walking and looking down at her cell phone when she entered the crosswalk. It was only a few seconds later she (Chelsea) looked up at the light. Once she saw the amber light she began to walk quickly.

Chelsea herself said that when she saw the light was amber she started jogging across. Therefore the reconstructionist considered a range of speeds from fast walking to average jogging speeds.

There are a multitude of tables of pedestrian speeds available from the research of Eubanks and Hill – walking, jogging and running, male and female, from the youngest to the oldest, with and without strollers, walkers, canes and shopping buggies, etc., and many combinations thereof. For many conditions, for example, 17-year-old females walking, values are given for the slowest end of the range (15th percentile – only 15% walk more slowly than this value), for the median (50th percentile), that is, where half the values are above the value and half are below, and for the fastest end of the range (85th percentile – only 15% walk faster than this value) (Eubanks & Hill, 1998).

In a legal report, I would consider a range of speeds all of which would be "normal". For the purposes of this chapter, and to simplify the analysis, I am only going to consider one value for Chelsea's speed. It is 7 km/h (4 mph), part way between fast walking speed and average jogging speed.

Unobstructed View Time: As Figure 6.1 shows, Mr Hartung's SUV, sitting in the left turn lane waiting for the light to change, obstructed Ms Abbas' view of Chelsea during her crossing of Main Street. In other words Ms Abbas did not have a line of sight to Chelsea, that is, she could not see her while Chelsea was crossing in front of Mr Hartung's SUV.

Figure 6.1 shows the area of the crosswalk obstructed due to the left turning SUV if Chelsea's speed was 7 km/h (4 mph) and Ms Abbas had been travelling at 25 km/h (16 mph) (as estimated by the reconstructionist) just prior to striking Chelsea.

To do a line-of-sight analysis and determine the obstruction due to the SUV, the reconstructionist needed to calculate the speed and position of each person as they approached the other on a second-by-second basis. Then the point at which the driver's view became obstructed and the point at which it became unobstructed again could be determined accurately. Reconstructionists often use simulation programmes to do these calculations. Putting Chelsea's estimated speed together with the Abbas Toyota speed, calculations show that once Chelsea appeared on the other side of the SUV, Ms Abbas would have had an unobstructed view of her for on the order of 1–2 seconds before impact.

Angle at Which Chelsea Was Seen in Peripheral Vision: Typically, the driver's line of sight is along the road path he or she is following. A driver whose attention is on the road ahead and, in this case, concentrating on hard-to-see traffic signals, will most likely be depending on peripheral vision to detect a crossing pedestrian. More on the challenges of that later.

The reconstructionist calculated the angle off the line of sight that Ms Abbas would see Chelsea, based on the speeds with which they were approaching each other. The more similar the speeds of the vehicle and the pedestrian, the larger the angle. If Chelsea had been moving at a fast walking speed (7.2 km/h (4.5 mph)), and Ms Abbas at 25 km/h (16 mph), Chelsea would have been seen at 16° off Ms Abbas' line of sight. As you will see from The Science section, the larger the angle off the driver's line of sight Chelsea was when she started across the intersection crosswalk, the less easily a mentally loaded driver approaching the intersection would notice her.

State of Traffic Signal: From her walking speed and the distance travelled, the reconstructionist determined that Chelsea had travelled south in the crosswalk for 9 seconds before she was struck. During this time, she would have been facing a green light for the first 2 seconds, an amber traffic light for the next 4 seconds, and a red light for 2 seconds (for both directions of traffic). In addition, we have to consider that if Ms Abbas entered the intersection after the light turned green, there would have been a slight delay between the light turning red and Ms Abbas entering the intersection. If we estimate that the delay is 1 second, then the total time Chelsea faced a red light was 3 seconds. Thus, Chelsea was struck 9 seconds after leaving the curb 8 seconds after the light had turned amber, and three seconds after the light had turned red.

The Investigation

The collision occurred in October at 8:00 a.m.

The sun was just rising and still at a low angle, in a position to cause significant glare, making it difficult for Ms Abbas to see the colour of the traffic light change. To have the same difficulty as for the driver, with regard to the lights, I visited the site under sunny conditions within a few days of the same time of year and the same time of day as the collision occurred. The timing of my visit ensured the sun was in the same position and created the same amount of glare as was the case when the crash occurred.

I was accompanied by my assistant, Ms Chen. Doing an accurate assessment requires that the pedestrian be dressed in similar colours, with similar reflectivity. Therefore, like Chelsea, Ms Chen was dressed in grey clothing. I was driving a vehicle with the same seated eye-height as the one driven by Ms Abbas. This ensured that when I looked straight ahead at the traffic signal, my line of sight was as close to the glaring sun as was the case for Ms Abbas. As it was at the time of the collision, the weather on the day of the site visit was sunny and dry.

There were two lanes in each direction on Main Street, and a left turn lane at the intersection with Larch Avenue. Based on police photographs of the scene of the collision and traffic light signal data, in addition to our own observation, nothing significant about the road design, traffic signs, street-lighting, etc. had changed at the intersection since the time of the collision. Checking for changes is important. Improvements such as brighter street-lighting or an additional right turn lane can affect your analysis. It can be very embarrassing, as happened to a colleague of mine, to do an analysis of the roadway as it existed at the time of the site visit without taking account of changes that were made between when the collision occurred and the site visit. Had my colleague's case gone to court, it would have called his whole analysis into question.

At the site visit, I wanted to see how difficult it would have been to see the traffic signal change from green to red when the rising sun was close by.

I wanted to get a sense of how many pedestrians there were at that time of day. During the 15 minute period before the time of the collision, that is, from 7:45 to 8:00 a.m. I observed only two pedestrians crossing Main Street at the traffic lights. This suggested to me that a driver familiar with this intersection might not be expecting pedestrians, especially crossing against a red light.

I also wanted to get a sense of driver workload related to the driving task. Using a pair of safety glasses with a mini camera between the eyes I videotaped the approach to the intersection with Larch Avenue following the path of Ms Abbas along Main Street. My assistant, dressed in grey clothing, stood in a slightly different location on each run. Movement makes a target in peripheral vision more noticeable. So when investigators view pedestrians, it is best that the target pedestrians move in place to help re-enact the situation the driver would have seen. Despite her movement, Ms Chen was not very conspicuous in her grey clothing seen against the grey pavement.

Before making any trial runs towards the intersection, I arranged with my assistant that I would drive past eight times. She was to stand at the same corner from which Chelsea attempted to cross the intersection, on four of the eight runs. Therefore, I did not know in advance whether my assistant would be present or not. By starting at a particular spot, when there was a mid-size vehicle or larger heading for the left-hand turn lane, we had the best chance of being able to observe and record the driver's path and the view blockage due to a vehicle in the left-hand turn lane.

It can be very time-consuming to obtain an accurate re-enactment, and in any case, safety concerns mean that it might not be advisable to do so. On one occasion of the eight trial runs there was a signal change when I was the first vehicle approaching the point of impact in the middle (passing lane) and there was a large vehicle (an SUV) in the left-hand turn lane. However, the light turned green for me too late and I had to stop before crossing the intersection because it was still red. Despite its flaws, which should be pointed out, an accurate re-enactment or simulation can be very helpful in leading members of the court to imagine themselves in the driver or the pedestrian's shoes.

On the four runs when she was present, I could see Ms Chen when she was standing on the far corner of the intersection. However, if she had continued to cross, and as I continued to cross, she would have been blocked from view by the vehicle in the left-turn lane. Because of the vehicle in the left lane, Ms Chen was very difficult to detect unless I deliberately turned my head to the left when I was close to the intersection. The driver must pay attention to many locations while approaching an intersection, and the driver in question was having difficulty seeing the colour of the traffic signal, due to the sun position, and was passing between vehicles on the right and on the left. A driver not expecting a pedestrian from the left, especially several seconds into the red light, would be unlikely to concentrate on the road scene to their left.

With respect to the driver's visual search task, I found it necessary to look straight ahead to monitor the traffic light and pass between the vehicle in the left turn lane and any traffic in the right lane. I had to trust that I would detect a pedestrian using my peripheral, or "out of the corner of my eye" vision. The closer I was to a signal change the more intently I had to watch the traffic light.

In the seconds before the collision, Ms Abbas was approaching the intersection on a red light that was about to turn green. A driver in this position would be likely to be looking at the path ahead through the intersection, stopped and moving vehicles ahead they were about to pass, the stop line, and the traffic signal. A driver who is not certain they will have to stop would be more engaged in the immediate question of whether they should stop or not than a driver that definitively knew he or she was going to have to stop. That decision was made more difficult by the glare due to the sun's close position to the traffic signal at that time of day and day of the year. Ms Abbas reported it was difficult to see the light change because of the proximity of the sun to the traffic light. I too found it difficult, necessitating squinting and staring hard at the light on the approach.

The Science

There are many human factors to be explored in relation to this collision. I have chosen four as being especially of interest:

- useful field of view (UFOV)
- mental load and useful field of view
- driver perception-reaction time (PRT)
- cell phone use and teenage pedestrians

Useful Field of View (UFOV)

While "field of view" generally refers to the total extent up and down, left and right that the eye can **see**, useful field of view (UFOV), refers to how far off the line of sight straight ahead (i.e. in the direction you are travelling) the eye is likely to **notice** things.

The UFOV is defined as the total visual field in which useful information can be acquired in a single glance. While your total field of view may be 90° to the left and 90° to the right, your UFOV will be much narrower depending on the difficulty of the task you are doing and the characteristics of the target to be detected (Wolfe et al., 2017) (See Figure 3.5 in Chapter 3).

At intersections, much is happening. Important sources of information are at larger angles from the driver's line of sight and therefore become more and more separated the closer the driver is to entering (see Figure 6.2). Examples of "sources of information" are the pedestrian on the far corner, the traffic light, and the stopped truck at the intersection ahead, etc.

In Figure 6.3 the circled area shows the UFOV a few seconds away from an intersection. It shows that, as the driver approaches the intersection, the UFOV covers less and less of the whole width of the intersection and right side of the road. A jogging pedestrian or a bicyclist can be on a collision course with a driver, and the driver fails to notice them because they are outside the driver's UFOV. The faster the pedestrian or bicyclist is moving, the less likely the driver is to notice them in time to stop, and the more dangerous the situation is. It should be noted that when two vehicles are on a collision course, the angle between them depends on their relative speeds. The more similar their speeds, the closer the angle between them is to 45°, well beyond the angle at which drivers are likely to detect peripheral targets.

Mental Load and Useful Field of View (UFOV)

Useful field of view is defined as the total visual field in which useful information can be acquired in a single eye fixation. UFOV is affected by how preoccupied or mentally busy you are with a central or primary task (Ball et al., 1988). In our case, the "central task" is driving. The primary tasks of driving involve controlling the path of a vehicle by steering, and its speed by braking. The secondary task is watching out for hazards. "Mental workload" is a more formal way of referring to preoccupation or busyness.

A laboratory study, by Ball et al. (1988), examined the effect of central task demand, number of distractors, and age on the ability to locate a target stimulus at 10°, 20°, or 30° off the line of sight during a single fixation. The greater the central demand, the greater the number of distractors and the greater the eccentricity, the more frequently targets were missed. Applying these findings to the accident in question, the more demanding the driving task and the further the target (the pedestrian) was off the line of sight, the longer it would take for the driver to notice the pedestrian.

FIGURE 6.2
(a, b, c) Sources of information are seen at increasingly large angles off the driver's line of sight as the driver approaches.

FIGURE 6.3
A fast-moving target can enter the UFOV from the side before a driver notices them in peripheral vision and stops.

On-road studies show that the laboratory findings about the increased difficulty of detecting peripheral targets apply to the real world as well. On-road studies have looked at the target properties and driving circumstances which affect target detection. In one on-road study, a group of driver subjects was asked to report whatever objects they **noticed** as they drove (e.g. sign, man standing at bus stop, mailbox, etc.) (Cole & Hughes, 1984). In other words, subjects were asked to report a "stream of consciousness" while driving.

Unbeknownst to the subjects, round disc targets, mounted on poles like traffic signs, had, prior to the study, been set up by the researchers. The locations were selected to allow the researchers to understand what made targets likely to be noticed (e.g. size, angle off the line of sight). Unlike the first group of subjects, a second group of driver subjects were told in advance of driving about the disc targets that had been set up along the roadway. This group of subjects was asked to search for the disc targets and report any that they noticed.

The drivers (second group) who were asked to search for the disc targets specifically, expected them and were considerably more likely to notice them (40% of targets detected) and report them as compared to the subjects (first group) who were not asked to search for them but just to report whatever they saw (only 6% of disc targets detected).

Ms Abbas' state of mind would be much more like that of the drivers in the first group, who were looking for and reporting on any signs, than they would be like the drivers in the second group, who were expecting and looking for specific signs – the disc targets.

When subjects drove in a busy urban area with shops (high mental workload), far fewer of the disc targets were reported (40% of targets detected) compared to driving in a peaceful residential area which was much less visually busy (low mental workload) (80% of targets detected). This is a big difference and shows how important workload is in how well drivers detect information in the driving environment. The shopping area would have had many more things to look at than the residential area, and not surprisingly, more disc targets were missed in the shopping area.

Ms Abbas had demanding central tasks (passing between stopped vehicles in adjacent lanes while watching the traffic light for a change to green, in the face of glare from the rising sun). In terms of her expectations about the presence of a pedestrian crossing on a red light, she would be like the subjects who were not given any information that they should expect sign disc targets.

Another important finding from the disc target study that applies to Ms Abbas' situation was that targets tended to first **be noticed** when they were near the line of sight, that is, when they were less than 10° off the driver's line of sight. The conclusion from these two on-road studies parallels the laboratory studies. The further a target is off a driver's line of sight and the more cluttered the environment, the less likely a target is to be detected and the longer it will take to notice. Targets most likely to be detected are less than 10° off the driver's line of sight.

Driver Perception-Reaction Time (PRT)

Dr Paul Olson, a well-known U.S. expert in forensic human factors, defined PRT as "an interval that starts when some object or reaction condition first becomes possible for a reasonably alert driver to detect, and ends when the driver has initiated a response (e.g. foot on the brake pedal or the hands start to turn the steering wheel)." Driving simulator studies are very convenient for getting measures of how drivers respond in emergency situations without taking any real risks.

A study using a driving simulator measured braking PRT to two unexpected pedestrians stepping off the curb and into the lane mid-block. The two pedestrians were visible for over 10 seconds on a number of occasions before the incident occurred. PRT was measured from when the pedestrians stepped into the lane of traffic until the driver touched the brake and **before any actual braking occurred**. Overall, PRTs ranged from about 1 to 1 ½ seconds (Broen & Chiang, 1996).

A Swedish on-road study had findings relevant to Ms Abbas' ability to detect Chelsea, measured the increase in PRT to the slowing of a vehicle ahead (detecting slowing was the secondary task). Drivers attended to displays within the vehicle (detecting changes in the displays was the central task). The displays were at various angles off the line of sight to the vehicle ahead (Lamble et al., 1999). Drivers were tested using a section of roadway closed to the public. Depending on the angle off the driver's line of sight, decreases in time to collision, that is, increases in time to detect a stimulus, were as much as 2.2 seconds over and above time to collision for stimuli seen on the line of sight. For an angle of 17°, this study would suggest an increase of 1.2 seconds over and above PRT involving a stimulus seen directly ahead.

Cell Phone Use and Teenage Pedestrians

Observations were made of teenager looking behaviour while crossing a road near a high school (Baswail et al, 2019). Approximately one third of crossings were made by teenagers carrying a phone or other device. They looked left and right before crossing less frequently when they had an electronic device with them, when looking at the screen and when texting or swiping. The authors concluded that teenage pedestrians were less safe when using mobile phones and music players while crossing a roadway.

Analysis

What does The science of human factors tell us about what a driver would be capable of in this situation?

First, let us consider a driver whose only chance to see Chelsea was when she, Chelsea that is, passed in front of the SUV in the left-hand turn lane and came into view. It would only have taken Chelsea 1 ½ seconds to get from in front of the SUV, where she was first clearly visible, to the middle of the driver's lane where she was struck. In the simulator study, described above, PRTs to a pedestrian stepping into the lane ranged from about 1 to 1 ½ seconds. The amount of time available for seeing Chelsea would have been enough for most drivers to get their foot on the brake, but not enough time to actually brake to a stop.

Any braking a driver can do to reduce the speed before striking a pedestrian is very worthwhile. At a striking speed as low as 30 km/h (19 mph), 5% of pedestrians are killed. At 60 km/h (38 mph), this rises to 60% being killed. Braking for even a fraction of a second helps.

Second, what about the driver's opportunity for seeing Chelsea **before** she passed in front of the SUV, between where she stepped off the curb corner and reached the SUV? During those few seconds a driver looking to their left could have seen her. But would it be likely? Here is where the research on what drivers can **see** versus what drivers **notice** comes into play.

Research that is described above has shown that targets, such as a pedestrian, are likely to be detected when they are close to the driver's line of sight, that is, within approximately 10 to 15°. In the situation in question, Chelsea was located 16° off the driver's line of sight. Under ideal circumstances (e.g.

low mental load with no obstructions, no difficulty seeing the traffic signal colour, no vehicles obstructing the view in the adjacent lanes), one would expect some drivers, perhaps most, would be able to spot Chelsea out of the corner of their eye, in other words in peripheral vision, and stop in time. However, the driver in this case was not operating under ideal circumstances at all. Even though several seconds were available **before** she was obstructed from view by the SUV many if not most drivers would **not** spot Chelsea in peripheral vision in time to stop. The driver was engaged in a difficult secondary task, with the sun beside the lights making the change in colour difficult to spot.

Depending on the combination of pedestrian and driver speeds, and individual driver PRT, a normally alert driver responding only after Chelsea became unobstructed by the SUV, might or might not have been able to brake enough to reduce the injury to Chelsea or to avoid the collision altogether. Whether the driver can reduce speed, and by how much, is a matter of fractions of a second. Some combinations of speed and PRT would give drivers enough time to perceive, react and bring the vehicle to a stop. Other combinations would result in lowered speed but not a stop. And some combinations would result in the driver hitting the pedestrian at speed and greater injury or death.

Conclusions

In forming an opinion on whether Ms Abbas did as well as could be expected from a reasonably cautious driver, I considered 1) the loading nature of the traffic signal task and the glare from the sun that interfered with that task, 2) the low probability that Chelsea would be detected in the few seconds from leaving the corner before she was obstructed by the Hartung SUV, and 3) the very brief time that Chelsea was unobstructed once she passed in front of the Hartung SUV. It was my opinion that given the circumstances Ms Abbas most likely had insufficient time to perceive and react to Chelsea, let alone to stop before hitting her.

A question often asked in court in a situation like this, where one cannot say for certain that a driver ought to have been able to stop is: on a balance of probabilities, is it more likely than not that fewer than 50% of reasonably cautious drivers would be able to stop in time to avoid hitting Chelsea. (Questions asked in court can sometimes be very complicated!) In this situation my answer would be yes, it is more likely than not that drivers would not be able to stop were they facing the challenges of Ms Abbas' situation. Under other circumstances or combination of circumstances, for example if there was no obstructing vehicle, or a much slower speed than 25 km/h (16 mph) allowing more time to react once Chelsea was unobstructed, my answer might have been "No, one cannot say that most drivers (i.e. more than 50%) would not be able to stop."

Although it was "stale" Chelsea did enter the crosswalk on a green light. The green turned to amber after 2 seconds. She was still in the crossing when the light turned red. There were five traffic lanes to cross and no pedestrian countdown timer or refuge island to assist pedestrians.

An independent witness said that Chelsea was looking at her phone while crossing. Chelsea could not remember if she looked at the traffic signal after she entered the crosswalk. She admitted she was carrying a cell phone when she was struck but insisted to the police that she had finished texting before she entered the crosswalk. It is less likely Chelsea would look for eastbound vehicles hidden by vehicles entering on the green phase.

Settlement

This case, like 95% of the cases in which I have given expert evidence, was settled out of court. The opposing parties would have, after negotiations, agreed on accepting a certain percentage of responsibility for the collision. It is seldom the case that one side is assigned 0% liability and the other 100%. It is also seldom the case that experts find out what proportion of the liability was assigned to their side of the case, or indeed even who won".

Countermeasures

A countermeasure that has been shown to be effective in reducing pedestrian-vehicle collisions at intersections is the Leading Pedestrian Interval (LPI). The LPI refers to a period of about 3–7 seconds, when pedestrians get a head start for crossing at a signal. By getting well into the crosswalk, before vehicles get the go-ahead (i.e. the green light), pedestrians can be seen by drivers entering just as their light turns green and collisions may be reduced by 13%. The LPI is an inexpensive but proven countermeasure (Goughnour et al., 2018).

References

Ball, K., Beard, B., Roenker, D., Miller, R. L., & Griggs, D. S. (1988). Age and visual search: Expanding the useful field of view. *Journal of the Optical Society of America, 35*, 2210–2219.

Baswail, A., Allison, L., Goddard, D., & Pfeffer, K. (2019). Adolescents' mobile phone use while crossing the road. *Safety, 5*(2), 27.

Broen, N. L., & Chiang, D. P. (1996). Braking response times for 100 drivers in the avoidance of an unexpected obstacle as measured in a driving simulator. In *Proceedings of the human factors and ergonomics society annual meeting* (Vol. 40, No. 18, pp. 900–904). Sage CA: Los Angeles, CA: SAGE Publications.

Cole, B. L., & Hughes, P. K. (1984). A field trial of attention and search conspicuity. *Human Factors, 26*(3), 299–313.

Eubanks, J. J., & Hill, P. L. (1998). *Pedestrian accident reconstruction and litigation.* Lawyers & Judges Publishing Co.

Goughnour, E., Carter, D., Lyon, C., Persaud, B., et al. (2018). *Safety evaluation of protected left turn phasing and leading pedestrian intervals on pedestrian safety.* Federal Highway Administration.

Lamble, D., Laakso, M., & Summala, H. (1999). Detection thresholds in car following situations and peripheral vision: Implications for positioning of visually demanding in-car displays. *Ergonomics, 42*(6), 807–815.

Wolfe, B., Dobres, J., Rosenholtz, R., & Reimer, B. (2017). More than the useful field: Considering peripheral vision in driving. *Applied Ergonomics, 65*, 316–325.

7

Disabled Vehicle No Lights vs Tractor-Trailer

Introduction

While driving on a highway at night in a rural area, a tractor-trailer driver suddenly encountered a disabled car with no lights, stopped on a narrow shoulder and protruding into his lane. The tractor trailer driver did not see the disabled vehicle in time to swerve or to brake. He struck the vehicle, resulting in injury to its two occupants. The human factors issues explored in The Science section deal with the visibility cues that were available to the truck driver and how far away a driver with normal vision would be expected to see the disabled vehicle. The Science section also explains the very large impact on seeing distance of expectation, that is, knowing that a hazard lies ahead.

The Collision

According to the police report, a crash involving a disabled passenger car and a tractor-trailer occurred near 1:00 a.m. in the winter on a highway through the mountains. The driver of the tractor-trailer was travelling at the speed limit of 90 km/h (56 mph) and shifting down his gears so he would not stall on an upgrade. As he was doing so, he suddenly encountered a passenger car stopped partially in his lane. He was unable to stop in time and struck the car forcefully. It was snowing lightly. At the time of the collision, the sky was dark, and there was no artificial lighting in the area.

A young couple, Brian and Anna Latham, had left a party at midnight in an old Chevrolet Impala they had borrowed from a friend. The engine had failed on the way there, but Mr Latham managed to get it going again. They hoped it would make the 50 km (31 mph) journey home without further problems. They took the main highway out of town.

As they continued it began to rain and shortly changed to snow. Then the engine started to cough and after a few minutes, failed, and refused to start again. This time the electrical system died. Mr Latham put the car in

DOI: 10.1201/9781003564805-7

neutral and the couple pushed the car backward off the road as far as they could. Because of a high snow-bank, they could only position the car so it was intruding into the lane by about half a lane width.

To stay out of the wind, both got back in the car. They had been in the car about 2 or 3 minutes when all of a sudden there was a strong jolt and a loud crash. They had been rear-ended by a tractor-trailer. They had no seatbelts on and both were badly injured.

The tractor-trailer was driven by John Churchill, aged 65. He said he saw the disabled car at the last second but not in time to even start braking. The police inquired how long Mr Churchill had been driving and from where he had come. This is standard practice when truck drivers are in a collision since a frequent cause of crashes involving trucks is the lack of sleep due to the demanding schedules that truck drivers keep. Mr Churchill said he had started that morning at 5:00 a.m. He had napped in the sleeper berth for 4 hours. His destination was about an hour away from the crash site.

The highway ran east and west and had two lanes in each direction (see Figure 7.1). Sheila Crummey was westbound in the passing lane passed by the disabled car partially blocking the curb lane. She told police that she passed a transport truck just before seeing a broken-down car partially blocking the curb lane. In her opinion, even if she had been in the curb lane herself, she would have seen the car in plenty of time to swerve around it had she been travelling in that lane. She did not see any lights on the car.

In this situation evidence from an independent witness such as this would be used to convince a judge and/or jury that the truck driver should also have been able to detect the broken-down car in time to move over. However,

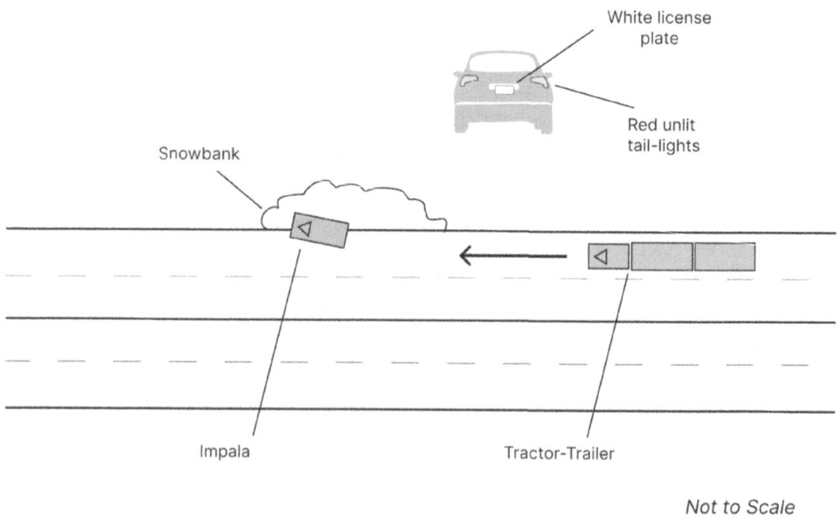

FIGURE 7.1
Disabled car struck by tractor trailer.

there can be many reasons that one driver sees a hazard in time and another does not. In this case, the witness driver approached in the passing lane. So she did not have to move out of the lane to avoid the broken-down car. She only estimated that she would have been able to avoid it. Her ability to avoid it was not tested.

The witness could have been more capable than Mr Churchill because of natural variability. People vary in physical characteristics such as height. For example, 5 ft 5′ (1.65 m) and 6 ft 2′ (1.88 m) are both normal heights for men. People also vary in perceptual and cognitive abilities. Drivers vary greatly, by as much as five to one, in how far away they will detect dark targets at night (Olson & Sivak, 1983 in Olson et al., 2010).

It is also important to note that this witness driver was 22 years old, and therefore likely to have better sensitivity to contrast, and therefore better seeing distance than the truck driver who, at 65 years old, was considerably older.

The Mandate

My job was to determine whether the disabled car would have been visible in enough time for a driver to avoid it, had he or she been driving in a normally cautious and alert fashion.

The Investigation

In the case of this collision, a site visit was necessary to assess how far away a normally alert and cautious driver could be expected to have seen the disabled car, partially blocking the lane nearest the shoulder without any lights, as it was at the time of the crash. It was critical that the assessment be done in a safe manner, without putting the investigators (myself and a reconstructionist), the truck driver, or the general public in harm's way.

The first order of business was to obtain a car that was a similar model and design to the disabled vehicle. Then we had to take it to the location of the crash where we met up with a tractor-trailer driver. Using his tractor, we intended to assess how far away the disabled car could be seen.

Once I knew the seeing distance, discussed below, I then determined adjustments that should be made to account for conditions on the night of the collision as compared to the night of the investigation. For example, because of being fully informed of the hazard that lay ahead, I was expecting it and knew exactly where and when to look for it. The tractor-trailer driver, of course, had no such warning to alert him. I had to account for the advantage of my state of awareness.

In order to avoid a crash, the seeing distance needed to be enough to account for distance travelled during two time periods. The first time period was the time from when the driver first perceived a potential hazard, recognized it as a hazard, and then initiated braking or swerving. This is referred to as perception-reaction time (PRT). It was my job to estimate it.

Let us say the higher end of PRT, based on various studies, is 2 seconds. At 70 km/h (43 mph), the truck would travel 20 m (65 ft) in each second or 40 m (131 ft) during 2 seconds.

But 40 m (131 ft) travel distance to account for PRT is not the end of the story. The driver still has to complete his or her braking or swerving motion to avoid the disabled vehicle. The second period is the distance travelled while braking or swerving is completed. The truck will travel some more distance while the driver steers it out of the lane with the protruding disabled vehicle or stops before striking it. It was the job of the reconstructionist to help determine how much time and therefore distance it would take to stop, as well as how much time and therefore distance it would take to swerve sufficiently away from the protruding vehicle to avoid it. It would depend, among other things, on the coefficient of friction of the road, that is, how slippery it was, the type of braking system (anti-lock or standard brakes), and so on.

Obtaining a Look-Alike Car

Typically, the car involved in the crash is no longer in a suitable condition to be used in the re-enactment of that crash. A look-alike or "exemplar car" must be obtained as a stand-in for the crash car (an Impala). This exemplar car must be as alike as possible to the crash car in the ways that matter for the issue of concern. In this case, that issue was the visibility of the disabled vehicle to other drivers. Therefore, we wanted a car with the same colour, or close to the same colour in terms of how much light was reflected by the tractor-trailer headlights, and the same reflectors, signal lights and type of licence plate on the rear of the car. Although there were no lights on the disabled vehicle, there were reflectors around the unlit taillights and side lights. These would help make the disabled vehicle visible.

The car involved in the crash was over 10 years old, making it very difficult to find a similar one. It took a month to find one to purchase.

Finding the right vehicle, just as finding the right location, can be very time consuming, and one has to be creative. In another case I had the challenge of finding an exemplar van, which had to be brown and over 12 years old. As I walked past a neighbour's house late one afternoon, carrying several bags of groceries, I saw the very van I needed parked on the street. Some ladies on a nearby porch told me where the owner lived. I plunked down my groceries on his doorstep, got my business card, rang the bell, and announced to the man who answered the door, that I would like to hire him to drive his vehicle on a nearby highway where a collision of the same colour and type of vehicle as his had occurred. I was a visibility expert and I wanted to make some observations. We wouldn't be doing anything unsafe. I offered an attractive sum of money. He was happy to agree. This method of locating a suitable vehicle was extremely efficient.

Environmental Conditions for the Re-Enactment

Generally, it is important to duplicate the environmental conditions for the re-enactment. If it was raining when the crash occurred, it would be preferable to do the re-enactment in rain. As you can imagine, it can be very difficult to organize rain, fog, and snow at the right time, and of the right amounts, for a site visit. In the case of a crash that requires you to undertake a long journey from home, as this one did, it is particularly difficult to arrange the investigation in the same weather conditions as were present at the time of the crash. In the case of this crash there was a snow-bank present, and the right side of the disabled vehicle would have been contrasted against it.

I considered what might be the effect of contrasting the disabled vehicle against a snow-bank on seeing distance. I reviewed the police photographs, attempting to find one that showed the snow-bank. Unfortunately, and not surprisingly these photographs had been taken several hours after the crash and it had been snowing the whole time since the crash. So they were not representative of what the surrounds looked like at the time of the crash. I decided I would try and roughly simulate the snow-bank at the site, using a large white cloth.

The Re-Enactment

I visited the site of the crash in the early spring, when there was still snow on the ground. I was accompanied by the accident reconstructionist, who was present to help carry out the required measurements. Based on weather reports we hoped to catch a late snowfall or at least a snow-bank left after winter snows. I drove a rental car. He drove what was to be the "exemplar" car, an old dark blue Impala like the one in the crash, that was purchased by the crash-involved trucking company for the purpose of the re-enactment. It was not in good condition – no windshield wiper fluid and a broken headlight we had to get replaced prior to leaving for the site.

It took us an hour and a half to reach the crash site. We drove at night in a mixture of intermittent rain, fog, and snow conditions. The lane marks on the highway were difficult to see. The reconstructionist asked me to drive behind him in case the exemplar car he was driving broke down. It was not the first time I had done a site visit with an exemplar car on the verge of a breakdown.

The weather was so bad that as I drove, I decided that if it was unchanged by the time the site visit was over, I would sleep in my rental car at the nearest "safe spot". I thought it would be safer than driving back in the poor weather in the dark. Fortunately, once we reached the crash site the rain stopped, and fog cleared. We had more fun on the return trip however. Our exemplar car was guzzling gas, and to avoid running out, the reconstructionist parked it at the one gas station on the return route, with a note explaining that someone would pick it up the next day.

While on a site visit, it is important to be prepared with a list of observations you want to make. In addition, there are fortuitous observations that can be made on the way to the site that are helpful. One of the things I look for is how far away I can see hazards similar to the one involved in the crash I am investigating. For example, in this case, which involved a passenger car without any lights, I was watching out for passenger cars without lights parked on the shoulder. I timed (by simply counting down the seconds (one one thousand, two one thousand, three one thousand, etc.)) how far away I could see them. I then converted the time into a distance by using my speed. So, for example, a car seen 2 seconds away while travelling at 90 km/h (56 mph) or 25 m/s (82 ft/sec), is detected at a distance of 50 m (2 seconds times 25 m/sec or 50 m distance (82 ft/sec or 164 ft distance)).

We had rented a tractor and hired a tractor-trailer driver to meet us at the site at 10:30 p.m. We chose that time because it was dark as it was at the time of the collision. Also we expected there to be little traffic interfering with our observations. The driver showed up as promised for which we were grateful. Nothing can be more frustrating than to go to great lengths to meet someone who you have hired to participate in a site visit, out in the middle of nowhere, and not find them there at the appointed time

As I already mentioned, the crash-involved passenger car was a green Chevrolet Impala. The Impala obtained for the re-construction was the same model, of similar colour (blue) with similar reflectance of light. The tractor the reconstructionist rented was a similar model to the tractor involved in the crash. We were now ready to proceed with the re-enactment.

We used the police report to determine the point of impact, which was about 40 m (131 ft) from a guide sign for the next town. For reasons of safety we did not place the Impala as it was at the time of the collision, partially blocking the lane. Instead we placed it on the shoulder, near but not over the edge line, and only for a minute or so for each trial run.

Because of the time it took to locate an exemplar vehicle and the lateness of the season, we were unable to arrange our visit under snow conditions. I tried to simulate the snow-bank using a large white cloth I had brought with me for that purpose. However, it was too difficult to arrange the cloth on the shoulder in such a manner as to get a satisfactory simulation and to do so safely.

Preparation for Test Runs

Before starting the test runs, the reconstructionist told the driver he would be parking the passenger car fully on the shoulder. He assured the truck driver that he would be safe as long as when he drove past the disabled

Impala, he maintained the tractor separation from the edge line of the right lane. The tractor driver and I waited in the tractor at a location about 1 minute travel time to the west of the crash site until there was **no traffic coming our way**. I told the tractor driver to place the tractor headlights on low beams as they were reported to have been at the time of crash.

I then alerted the reconstructionist. When there was **no traffic approaching from behind**, he turned all the lights of the Impala off (as drivers who witnessed the disabled vehicle indicated to be the case). He stepped well away from the car and gave a go-ahead signal to me via cell phone.

According to the reconstructionist, the speed recorded on the electronic data recorder (EDR) in the tractor-trailer just prior to impact was 70 km/h (43 mph). As an aside, many drivers, maybe even most, are unaware that their car, if it has an airbag, also has an EDR. The EDR records speed, acceleration, brake application, and more. The important thing is that the EDR retains the last 5 seconds of data. If an impact causes the airbag to be activated, the last 5 seconds is saved. The EDR data makes for very reliable evidence about speed, when the brake was applied etc. It can put the lie to a driver who claims, after a crash, to have been driving the speed limit when the EDR shows that he or she was actually going much faster when the crash occurred.

Test Runs

As noted above, we had placed the Impala **fully on the shoulder** with its wheels on the edge line at the impact location. Once I received the all clear from the reconstructionist, I instructed the driver to quickly bring his vehicle up to 70 km/h (44 mph) or almost 20 m/s (66 ft/s). He was to drive towards the Impala and pass it without slowing, if he felt it safe to do so.

A series of four test runs were made in the tractor towards the Chevrolet Impala. On each run I sat in the passenger seat of the tractor. I used a push-button on a device, provided by the reconstructionist, to indicate when:

1) I could first see the Chevrolet Impala
2) the front of the tractor passed the Chevrolet Impala

The difference between these two times gave me the number of seconds away at which an approaching driver could see the disabled car. We were travelling at 70 km/h (44 mph), or almost 20 m (66 ft) every second. At this speed, a 5 second travel distance is equivalent to 100 m (328 ft).

On each run I wore a headset with a mini-camera, showing the view from the tractor passenger seat. The headset also recorded my comments on visibility, and, in particular, 1) and 2) above. The naked eye can see further than a video camera. Therefore, it is more accurate to use the audio recording on the video to measure seconds of visibility to the target (the Impala) rather than deducing the time of seeing the target from the video image. While the video image gives a good general impression of the scene, it cannot be used to accurately measure seeing distance. This is because one can see low contrast targets further down the road with the naked eye than one can see with the video camera.

In a re-enactment, it often takes several runs before one obtains a run where everything goes to plan. However, one can learn something from each run. This is what happened.

On Run 1, due to a miscommunication with the reconstructionist, the lights of the Chevrolet Impala were on and the reconstructionist was still in the car as the truck driver approached at 70 km/h (44 mph). I was looking for a parked car on the shoulder without lights. I was not expecting a car on the shoulder with lights. The video indicated that, the **lit taillights** were in view on the video for about 23 seconds. However, a total of 13 seconds elapsed before I reported seeing them. The fact that I was so delayed in recognizing what I was looking at is a powerful example of the effect of expectations on detection distance to a hazard.

On Run 2, the taillights of the Impala were off, as they were supposed to be. As I expected, the most prominent cue by far as to the presence of the disabled vehicle was the white licence plate contrasted against the dark car body, followed quickly by the two dull red unlit taillights, these three elements creating the familiar triangular pattern seen on the rear end of vehicles.

Based on my verbal report, recorded on the video, of first seeing the licence plate to when we passed it, I could see it about 6 seconds away at 70 km/h (44 mph). The 6 second measurement was determined by timing with a stop-watch the delay between when I verbally reported seeing the Impala and when I verbally reported passing the Impala, as recorded on the videotape. I also timed the delay between my verbal report of seeing the Impala and its appearance on the videotape. There was a 3 second delay. In other words, the naked eye saw further than the video camera.

At this point we decided to look at photographs of crash-involved car to make sure it showed similar visibility to the exemplar car. (We should have done this earlier.) It did not. The licence plates looked dirtier than those of the exemplar Impala. That was one thing. But just as important, and even more critical to visibility, the reconstructionist noticed that the exemplar car was equipped with licence plates which had a "retroreflective surface". (The exemplar car and the rental car were from different provinces. Provinces have different requirements for retroreflective licence plates.) "Retroreflective" refers to the sheeting used which reflects the headlights of a car, channelling it so that a maximum amount of light is reflected back to the driver's eye. Thus the "exemplar car" being used in the re-enactment had more visible plates than the car involved in the crash. This would constitute a bias against our client, the trucking company, if it was not corrected.

Where can one get a non-retroreflectorized licence plate at 11:00 p.m. in the middle of nowhere? Fortunately, the reconstructionist had a screwdriver and had the brilliant idea to remove the rear non-reflectorized licence plate from my rental car and mount it on the back of the exemplar Impala. The Impala was placed back on the shoulder of the roadway, now with the appropriately less visible rear licence plate.

On Run 3, with the Chevrolet Impala equipped with a more representative and less visible rear plate and with its headlights and taillights off, we were hoping for a final successful run. Perhaps third time a charm.

Unfortunately, no! On our third approach to the Impala, the truck driver, to my surprise, suddenly flashed his high beams. This enabled me to see the disabled car earlier than I would have, had the lights been on low beam. He knew I would not be happy about this and immediately explained why he did so. He knew from our position, close to a guide sign, that we were almost at the Impala, with its wheels on the edge line, and **he could not see it**. Rather than risk striking it, he invalidated the test by flashing his lights when we were 7 seconds from the Chevrolet Impala. Note, this was despite prior instructions that he would be safe as long as he maintained the tractor separation from the edge line of the right lane when driving past the Impala. His response, given he could not see the Impala, was certainly appropriate. With the assistance of the (unplanned) high beams I could see the Impala at about 6 seconds' travel distance at 70 km/h (44 mph) according to the reconstructionist's measurement device.

Before attempting Run 4, the reconstructionist reassured the tractor driver again that because of the car was fully on the shoulder, as long as he maintained separation from the edge line he would be fine.

On Run 4, with the Chevrolet Impala equipped with a rear non-reflective white licence plate, its lights off, and the tractor driver reassured on the important matter of safety, I could see the exemplar Impala about 2 seconds (timing from voice on video) away at 70 km/h (42 mph). Finally, we had a reasonably accurate re-enactment run!

In summary, the four runs showed:

1) **Lit taillights could be** seen on video at the equivalent of **23 seconds'** travel distance to the Impala. (They were likely visible earlier, as the naked eye can see further than indicated by a video image.) This measurement shows the distance at which the disabled car might have been detected had the taillights been functioning (**Run 1**).
2) When the taillights were off, the most prominent cue that there was a vehicle near the lane edge was the white licence plate, followed by the unlit but still reflective red taillights. The licence plate that was used on the exemplar vehicle was less reflective than the one that had been on the disabled car at the time of the collision (**Run 2**).
3) The retroreflective licence plate and **unlit taillights** were seen by a passenger (me) expecting to see them at the equivalent of **6 seconds'** travel distance, considerably less than lit taillights would have

been seen. Seeing distance was overestimated due to the presence of brighter (retroreflective) licence plates being used than were on the disabled car at the time of the crash, high beam lights being briefly flashed on near the disabled vehicle, and no correction being made for driver expectancy (**Run 2**).

4) The target (licence plate) **reported on the video's** audio recording could be seen with my naked eye 3.0 seconds earlier than it was **visible on the video** (**Run 2**).
5) The non-retroreflective licence plate and unlit taillights were seen at approximately 6 seconds with brief assistance from high beams flashed at 7 seconds away (**Run 3**).
6) There were miscellaneous white lights (both headlights and house lights) and white reflections (post-mounted delineators, signs) at the site, as on all highways at night. These could slow the detection of the triangular pattern for other drivers, as they did for me. The non-retroreflective licence plate and unlit taillights were seen at approximately 2 seconds. The observer (myself) was delayed by initially attending to a close-by white light (**Run 4**).

In summary, based on a re-enactment with a similar passenger car, with unlit headlights, with the non-retroreflective white licence plate, seeing distance to the disabled car that was struck in the crash would have been short – approximately 2–6 seconds' travel distance. At 70 km/h (44 mph) or about 20 m (66 ft) per second, seeing distance is about 40–120 m (131–393 ft). This, of course, is for an observer (myself) who had brief assistance from high beams flashed at 7 seconds away, who was very alerted, and knew approximately where the exemplar car was. More on that later.

The Science

The re-enactment provided much useful information about cues, which a driver such as Mr Churchill would have had to use to detect the disabled vehicle. The cues indicating that a vehicle was present and blocking the roadway would have come first from the non-retroreflective white licence plate, second from the retroreflective portions of the red taillights, and third from light reflected by the rear-facing portion of the car body.

Where the re-enactment falls short is in the number of subjects used to estimate seeing distance. There were two runs with a non-retroreflective licence plate with one older subject – myself. This estimate needs to be backed up if possible, by research studies of visibility distances with low beam headlights to targets similar to the target in question involving a wide range of subjects. Fortunately, there are such studies.

Seeing Distance

Dr Paul Olson and his colleagues at the University of Michigan Transportation Research Institute carried out an extensive experiment in which driver seeing distances with low beam headlights were measured for 23 subject drivers, various targets and situations (Olson & Sivak, 1983 in Olson et al., 2010). These included pedestrians as well as rectangular shapes of various sizes. The pedestrians were in dark clothing or had white vests. They stood on the right side of the road where low beam headlights are aimed to avoid glare for the oncoming driver, as well as on the left side where visibility is half that of targets on the right side. Seeing distance was measured by having passengers in an approaching car push a button as soon as they could see the target.

One issue which must be mentioned is that headlight beam patterns, both low beam and high beam, have undergone some changes since these results were collected. In this case the tractor-trailer driver was using his low beams. By far the majority of the time drivers are using their low beams. This is to avoid constant switching back and forth between high and low beams. The main power of the low beam is directed down and to the right. This is to avoid creating glare for oncoming drivers, but it is at the cost of having less seeing distance.

The low beam headlight beam pattern (how much light is aimed where) in Canada and the United States is dictated by a Society of Automotive Engineers (SAE) standard. It is this and not the lighting technology that determines the seeing distance. All low beam systems provide approximately the same visibility, and it is very limited when compared to high beam headlights (Olson et al., 2010).

Which of these test targets and conditions was most similar that of the collision in question? The re-enactment showed that the most noticeable part of the disabled car was the white licence plate. The closest object to that white licence plate and the white snow-bank, in terms of reflection of light, and therefore seeing distance, would have been the pedestrian in the non-retroreflective white vest standing on the right side of the road. On the one hand a vest is larger than a licence plate. This would make the vest more visible than the licence plate. However, the vest would be seen higher off the road, where there is less light from headlights that are pointed down. This would make the vest less visible. The range of seeing distance for the white vest on the right side of road was approximately 50–140 m (164–459 ft).

This range (50 m–140 m (164–459 ft)) compares well with the 40–120 m (131–393 ft) (2–6 seconds) estimate of seeing distance from the re-enactment. It should be noted that older subjects detected low-contrast targets at an average of about half the distance as did younger subjects. In other words, older drivers would get much closer before they detected the disabled vehicle. Mr Churchill was 65 years old at the time of the collision.

Driver Expectancy

A stopped car partially blocking the lane on a high-speed limit highway is a most unusual occurrence. Drivers are likely to drive for many years without running into such a situation. In my 40 years of driving I have, thankfully, only encountered that once. I was travelling in the middle lane of a five lane per direction highway at night, keeping up with traffic at 110 km/h (68 mph) or so, when I suddenly encountered a tractor-trailer without lights stopped in my lane. There was no time to check adjacent lanes for traffic; I simply swerved into the next lane and was very fortunate there was no one in the way. Similarly, Mr Churchill would be unlikely to have encountered a disabled vehicle, not on the shoulder, but protruding substantially into the lane very often.

The difference expectation can make in the detection of targets at night is illustrated by a clever study of pedestrian visibility performed in the United States (Roper & Howard, 1938). In this study drivers were recruited under the pretence that their task was to assess the quality of headlights. After some time, enough to become familiar with the car, they were asked for their assessment and then directed home along an unlit rural road. On the way home, feeling relaxed, the driver was suddenly confronted by a pedestrian target on the road. The invariable reaction was first to release the accelerator. This started an odometer which was stopped when the driver reached the pedestrian target. Once the driver knew exactly where the target was and what it looked like, he was asked to back the car up, drive towards it, and again release the accelerator when he could just see the target. **On average**, response distances obtained the second time were **twice** those obtained when the driver was surprised. There are several reasons why the second recorded seeing distance was so much longer. Knowing that the target will appear on the road, that it will be in particular position (e.g. on the right side of the road, just after a sign showing the speed limit on the left side), allowed the drivers to focus their attention to where the target first appeared. Knowing about subtle cues such as light-coloured shoes or the approximate height of the target, would allow drivers to locate the pedestrian target more quickly. The results of this study were subsequently corroborated by Shinar (1985), and cited in the highly regarded book *Forensic Aspects of Perception and Response* (2nd Edition) (Olson & Farber, 2003).

Consequently, visibility distances measured in experiments or in reconstructions, as was done in this case, need to be corrected to account for expectancy by assuming that the seeing distance on average in real life is 50% of the seeing distance measured in an experiment. If this correction factor is applied to the seeing distances above, the revised estimate for the normal range of seeing distances for young drivers not expecting to see a pedestrian in a white vest, standing in for the white licence plate on the disabled vehicle, seen on the right side of the road, is 25–70 m (82–230 ft). Similarly, the estimate of seeing distance (2–6 seconds' travel distance at 70 km/h (44 mph)) based on the on-site study would be 50% of the seeing distance in the real world, where drivers have no warning of a hazardous situation, such as the unlit disabled vehicle partially on the road.

Mr Churchill needed to do more than see and recognize the disabled vehicle. He needed to move his vehicle into the adjacent lane to go around it or he needed to stop his vehicle before he reached the car.

A study in Sweden measured how long it took for truck drivers to steer away from a light at the side of a rural road at night which suddenly illuminated. To move a small amount over in the lane took 2.5–3.25 seconds (Summala 1981). A bigger swerve, taking longer, would be required to avoid the disabled Impala, which was protruding about one third of the way into the lane according to the reconstructionist. At 70 km/h (44 mph) the tractor-trailer would travel about 65 m (213 ft) during 3.25 seconds. Thus, normally alert drivers with short seeing distances and/or slow swerving times will crash!

With regard to the effect of the snow-bank on visibility, I considered that the major cue as to the presence of the unlit vehicle was the white licence plate. It would have been seen contrasted with the dark car body. The white licence plate was nearer the driver's line of sight than the snow-bank and so would attract the driver's attention sooner. The snow-bank, like the licence plate, was not retroreflective. I concluded that the seeing distance measured at our site visit would probably provide a reasonably accurate estimate of the seeing distance to the disabled Impala on the night of the crash. To be absolutely sure it would be necessary to re-enact the situation with a snow-bank such as existed at the time of the crash, a difficult challenge to meet.

Other Issues

As I noted in the introduction, just some of the scientific issues considered in each case would be discussed. That is true here as well. For example, other issues that could have been discussed in this chapter include the effects of pattern recognition, visual clutter and age on seeing distance. I have often said a book could be written on each of these cases, there is so much revealed about human behaviour in each one.

Conclusion

As Mr Churchill approached the Chevrolet Impala at night, there were no flashers or lights warning of the partial blockage of a live lane by a disabled vehicle. The most noticeable cue to its presence would have been the triangular pattern created by the white licence plate and unlit red taillight reflectors seen against the dark car body. An assessment using a truck and a car with a similar rear end at the site showed two things; first, seeing distance to the disabled Impala would have been insufficient for many drivers to stop or swerve in time. Second, recognition of the triangular pattern created by the white licence plate and the unlit red taillights on Brian Latham's Impala could have been delayed as Mr Churchill was distracted by the presence of other lights and reflectors in the vicinity, not on disabled vehicles.

IS AVOIDANCE POSSIBLE?

Seeing Distance Under Ideal Conditions

50–140 m (164–459 ft) for **expected** hazard

Take half

25–70 m (82–230 ft) for **unexpected** hazard

Travel Distance During Perception, Reaction and Swerving:

2.5–3.25 seconds:

50–65 m (164–213 ft)

Drivers with short seeing distances and/or slow swerving times will crash!

In forming my opinion about Mr Churchill's performance, I considered:

- the number of seconds away (2–6 seconds) I could see the most visible cue on the disabled Impala, the white licence plate
- experimental results for a range of typical drivers to a small white target seen at night
- a number of other factors not discussed here, including the impact of visual clutter created by the other retroreflective devices near the roadway, which may have distracted Mr Churchill for a critical second or two, as they did me, as I looked ahead at the road
- and most importantly the fact that, unlike myself and unlike the drivers in the visibility experiment led by Dr Olson, that Mr Churchill had no warning of the presence of a disabled vehicle ahead

Based on these considerations, in the end, it was my opinion that about half of normal, reasonably alert drivers who were not expecting to encounter a vehicle without lights, partially in the lane, would be able to avoid a collision. The other half would not be able to stop or swerve in time.

Countermeasures

One countermeasure would be retroreflective licence plates. We were unable to find any methodologically acceptable (e.g. with big enough sample size) studies comparing them in terms of crash rate or even visibility distance to non-retroreflective licence plates. Before recommending these it would be important to know that they were effective.

References

Olson, P. L., Dewar, R. E., & Farber, E. (2010). The visibility provided by vehicle lighting systems, expectancy. In *Forensic aspects of driver perception and response* (3rd ed., pp. 175–211). Lawyers & Judges Publishing Company.

Olson, P. L., & Farber, E. (2003). Evaluation of visibility in the field. In *Forensic aspects of driver perception and response* (2nd ed., pp. 257–275). Lawyers & Judges Publishing Company.

Roper, V. J., & Howard, E. A. (1938). Seeing with motorcar headlamps. *Illumination Engineering, 33*, 412–438.

Shinar, D. (1985). The effects of expectancy, clothing reflectance, and detection criterion on nighttime pedestrian visibility. *Human Factors, 27*(3), 327–333.

Summala, H. (1981). Drivers' steering reaction to a light stimulus on a dark road. *Ergonomics, 24*(2), 125–131.

8

Bus vs Train – Fatigue?

Introduction

This chapter describes a collision at a railway crossing in which a bus driver drove into the side of a train, killing himself and five bus passengers, and seriously injuring nine others. Given the seriousness of this crash involving the general public and resulting in multiple injuries and deaths, it was investigated by the Transportation Safety Board of Canada (TSB) — an independent agency that investigates air, rail, marine, and pipeline occurrences (and selected roadway crashes when they involve one of the other modes of transportation). It is not the function of the TSB to find fault. It is rather to report fully on the causes and contributing factors to an incident so that a similar event will be less likely in future.

There were many possible causes of this crash. One was "driver fatigue". Because of my expertise in driver fatigue and its effects on performance, I was asked by the TSB to thoroughly review the evidence and The Science concerning fatigue to determine if it could have been a factor for the bus driver in this collision.

As you will see, whether or not "fatigue" is present is not a simple question that can be checked off by a police officer on a motor vehicle collision report. The Science section of this chapter deals with what is meant by "fatigue", as well as the many factors which can lead to fatigue's effects, including lowered alertness and poor performance.

The Collision

A passenger train, while proceeding west in Ottawa, Ontario, was struck by a double-decker city bus at a railway crossing. The crash happened on a Wednesday morning on September 18th, 2013, just before 9:00 a.m. The location was a roadway designated for buses only, called a transitway, at a point just past a transit station. The crossing was equipped with flashing light

 DOI: 10.1201/9781003564805-8

signals, bells, and gates that would be activated when a train was nearby or at the crossing. At the time of the collision, the flashing light signals, bells, and gates were operating as intended. **The gates had been down for at least 26 seconds prior to the collision.**

As the bus approached the crossing, there were no other buses on the transitway. Had a lead vehicle been present and slowing for or stopped at the crossing, the driver might have been alerted to the need to do likewise.

Figures 8.1 and 8.2 show the rail crossing at which the collision occurred and the resulting damage to the bus.

According to the TSB report on the incident, the bus was initially travelling at a speed of 68 km/h (43 mph), in a speed limit zone of 60 km/h. The driver began to apply the brakes when the bus was approximately 2–3 seconds from striking the train. He managed to reduce his speed considerably but was still travelling about 8 km/h (5 mph) when the collision occurred. Considering the slow speed of the bus, the cost to human life and the number of serious injuries was considerable.

As for the train, the train crew applied the emergency brakes 2 seconds prior to the collision. It can take over 40 seconds for a passenger train to stop. It would not be reasonable to expect the train engineer to detect, in time to avoid a crash, the highly unusual circumstance in which a bus loaded with passengers was not going to stop for lowered gates and flashing warning lights at a crossing.

TSB's investigation identified the main causes of the collision. My own investigation dealt with the possibility that fatigue may have been one of those causes. That is the subject of this chapter. To complete the story, at the

FIGURE 8.1
Rail crossing at which the crash occurred.

FIGURE 8.2
Damage to the double-decker bus. From TSB Railway Investigation Report R13T0192.

end of the chapter I will briefly describe the TSB's main conclusions as to the contributing factors to the crash.

The Mandate

My mandate was to assist the TSB by thoroughly reviewing the evidence and The Science concerning fatigue to determine if it could have been a factor in this collision.

The Science

In this section on The Science of fatigue, I briefly consider each of a number of possible factors which might have influenced the bus driver's level of fatigue, alertness, and performance. These factors are:

- recent sleep disruption
- long term sleep disruption

- duration of continuous wakefulness at time of the crash
- time of day (circadian rhythm) effects
- work-rest schedule
- sleep disorders
- medical and psychological conditions

The identification of these factors is based on my own knowledge of fatigue studies. In addition, I found the TSB's internal guide to investigating sleep-related fatigue, summarized in Rudin-Brown & Rosberg (2021), very useful.

Many of the factors listed above required information from the driver himself. In this case, the bus driver did not survive the collision, so I had to fall back on information that was provided by the TSB, taken from interviews with his family, co-workers, and the bus company.

Fatigue Analysis

Recent Sleep Disruption: Last 3 Days

The first indicator of fatigue that comes to mind is getting poor quality or quantity of sleep over the last few days. To decide whether or not insufficient sleep during the nights prior to the crash **was** a contributing factor, I needed to calculate how much sleep the driver had had in the week prior to the crash. Of course, I did not have any recorded information that would give me precise sleep times. Instead, I had to make a rough estimate using information provided to me by the TSB about:

- the driver's regular sleep habits
- his work schedule, and
- time between awakening and starting his shift

I also used what is known about normal sleep durations in the population.

Our driver went to bed typically around 10:00 p.m., 11:00 p.m. at the latest. The time depended on when he had to get up in the morning. In estimating how much sleep he had had, I assumed 10:30 p.m. for the start of sleep. In the absence of a reliable estimate of the driver's sleeping time, I assumed a normal pattern of 7–8 hours.

Based on information provided by the TSB, I also assumed that each day the driver got up 1.25 hours prior to the scheduled run time for his bus. This allowed for 52 minutes for him to shower, to dress, and to eat breakfast, to commute to work (estimated using Google Maps), and to arrive 15 minutes prior to the scheduled run time as required. Using the information outlined above, I estimated wake-up and bedtimes, and from them, how long the driver had slept.

Sleep Quantity: Last Three Days

The driver's last main sleep period prior to the morning of the crash was estimated to be 6.6 hours. On two of the three prior days the driver had short sleep – 6.6 and 6.2 hours. This is low compared to a typical length in the general population of 7–8 hours.

Even small reductions in sleep can affect performance and crash risk. An example with which many of you are no doubt familiar occurs when the clocks change for daylight savings time. The hour of sleep lost in the spring due to the time change has been found to lead to a measurable increase in crash risk (Coate & Markowitz, 2004).

A driving simulator study looked at the situation of our driver by measuring how long it took driving performance (e.g. staying centred in the lane) to get back to normal after driving during the daytime, the subject having had a short sleep the night before (Balkin et al, 2000). It found that the effect of sleep duration on performance was significant and depended on how short the night sleep had been. Remarkably, this effect was even seen for participants who were only mildly sleep restricted – to 7 hours instead of 8 hours in bed during the night before the work period. It probably seems surprising to some of you that a difference in performance between 7 and 8 hours of sleep is detectable. After all, 7 hours is close to what would be considered by many as a full sleep period.

The impact of the driver's short sleep would be lessened by the fact that he had only been awake for 3.7 hours when the incident occurred and by fact that the collision took place in the morning (more on that later).

Sleep Quality: Last 3 Days

With respect to sleep quality, as opposed to sleep quantity, the driver was often concerned that he might not wake up when his alarm went off and he deliberately woke himself up to check the time.

Another quality-related sleep disturbance might have been due to his inadvertently drinking caffeinated tea before bed. One cup of coffee improves alertness for about 1–3 hours. The driver typically had a cup of chai tea, with about one third the caffeine of one cup of coffee, before bed. These two factors (waking up several times and drinking a caffeinated beverage) would tend to reduce the driver's quality of sleep. On the other hand, the driver's sleeping environment was generally good (e.g. no young children at home), a positive factor.

Our driver was a good sleeper; he liked to sleep and found it easy to fall asleep and to sleep in on a day off. He worked split shifts and would often come home from work and take naps. He often napped between shifts on

Wednesday, Thursday, and Friday. Working split shifts has been shown to increase fatigue and its effects on performance (Anund et al., 2018).

I concluded that, based on two out of three prior days with short sleep and a minor concern about the quality of the sleep, a fatigue risk may have existed related to recent sleep disruption.

Long Term Sleep Disruption

Long term sleep disruption refers to reductions in the **quality** or **quantity** of sleep that have been experienced **for longer** than 3 days. The term "sleep debt" is used for the difference between the number of hours actually slept and the estimated 7–8 hours required, that is, "normal" sleep. The TSB uses a method for estimating a person's sleep debt by comparing the amount of sleep obtained to the amount of sleep required. There is a balance that should be maintained between the sleep needed in a 24-hour period and the number of work hours.

How does shiftwork affect sleeping times? A fair amount, it seems. A study of power station workers found average sleeping times of about 6 hours after a night shift, 7 ½ hours after a late day shift, and 6 ¾ hours after an early morning shift. The poorest sleep was found after night shift and the second poorest sleep after early morning shifts, such as the one that our driver worked. These shifts are typically associated with some sleep debt (Pternitis, 1977; Grandjean, 1982).

Our driver was typically awakened by an alarm. This, and the fact that he could sleep until 10:30 a.m. on a day off, suggests that his morning sleep was interrupted by rising between 4:00 and 5:45 a.m. as he needed to on days he was working, instead of a more desirable 5:30 to 6:30 a.m.

Sleep debt builds up over the work week. The TSB calculates sleep debt by considering the total sleep time, the total awake time, and the desirable ratio of sleep to wakefulness. An ideal ratio would be 1 hour of sleep for every 2 hours of wakefulness, or 8 hours sleep to 16 hours of wakefulness.

The collision occurred on Wednesday. The prior Friday and Saturday night sleeps would have allowed full recovery from any prior sleep debt. Thus, the calculation of the driver's sleep debt that existed at the time of the collision is started with the sleep which ended at 10:30 a.m. on Sunday and ends with the awake time prior to the crash.

The driver's sleep debt at the time of the collision was 1.3 hours. Once the sleep debt reaches 8 hours, then fatigue is considered to be a contributing factor. But even if it is less than 8 hours but greater than 0, as in this case, then fatigue may still have been a contributing factor.

Sleep Required to Recover from a Sleep Debt

A concern for commercial drivers is the amount of time needed for recovery from a sleep debt after multiple consecutive days' work. For example,

Canadian hours of service regulations require that, once a driver has worked 70 hours in a 7-day work cycle, they are required to have a period of off-duty time of at least 36 consecutive hours (Canadian Hours of Service). For drivers who work a 14-day cycle, the requirement is for an off-duty period of at least 72 hours after accumulation of 120 hours of work. These requirements provide for a minimum period of two nights of sleep between cycles. Many of you will know from your own experience that night sleep is more restorative than sleep taken during the day. Based on driver sleep and lane tracking performance, a study showed that a time-off period which allowed two full nights and 1 full day off allowed full recovery from daytime driving sleep debt (O'Neill et al., 1999).

The last 2 days in a row our driver had off were September 14th (sick day) and September 15th (day off). Although on September 14th he awoke to report in sick around 6:00 a.m., he went back to bed and slept late. These 2 days off would have allowed him two full nights of sleep, which studies indicate should be enough for full recovery from any sleep debt accumulated during his previous work week.

I concluded that, based on two nights recovery near the end of his last week worked before the crash, and a sleep debt of only 1.3 hours, a fatigue risk did not exist due to long term sleep disruption.

Continuous Wakefulness

Research has shown that 22 hours of continuous wakefulness may be considered the point at which fatigue causes almost all aspects of human performance to start to decline and micro-sleeps to begin. Micro-sleeps are very brief sleeps of a few seconds. A driver may be completely unaware that they are occurring. Drivers who fall asleep for a few seconds may be perfectly capable of continuing on the roadway but will be more likely to miss or to have a delayed response to sudden hazards.

In the case of the bus driver, he had been continuously awake for only 3.7 hours at the time of the collision, considerably less than the 22-hour threshold when micro-sleeps begin. Furthermore, his time continuously awake had occurred within a couple of hours of night-time sleep, when he would be most refreshed.

I concluded that continuous wakefulness at the time of the collision was only 3.7 hours and was not a factor in this crash.

Circadian or Body Clock Effects

Alertness and its counterpart, sleepiness, are determined by our body clocks or circadian rhythm. This is a daily variation of human physiological functions such as sleep, body temperature, blood pressure, adrenalin production, and many others. The circadian rhythm physiologically gears the body for action during the day, and for sleep at night. Body temperature, for example, rises slowly throughout the day, peaking at about 8:00 p.m. It then falls rapidly to reach a low point at approximately 3:00 a.m. (Grandjean, 1982). You may have become aware of this if you have been up at 3:00 a.m. and felt chilled.

Other physical functions have similar but not identical rhythms. Mental functions also have a 24-hour cycle. Mental abilities peak during the daylight hours. They are poorest at night, particularly around 3:00 am. There is a secondary low point in the early afternoon, known as the "post-lunch dip".

Individuals vary by as much as 2 ½ hours in the time at which body functions and mental abilities reach a low point. (Individuals whose circadian rhythms peak earlier are referred to as "larks". Those whose rhythms peak later are referred to as "owls".) As alertness declines, sleepiness increases. Various studies from many countries show a very large increase in crash risk by time of day, with the early morning hours having as much as 25 times higher crash risk per kilometre driven than other hours of the day.

When one works at night on a regular basis, the circadian rhythm can be shifted so that the body is more geared for action at night and for sleep during the day. But this takes a week or more to develop and adaptation remains incomplete because the cues that synchronize the body clock with the 24-hour day, such as the rising and setting of the sun, are continually in conflict with other cues, such as work hours.

Because of this incomplete adaptation, referred to as "circadian desynchronization", performance is not optimal and night-shift workers are vulnerable to motor vehicle crashes. For example, a survey of shift-working emergency medicine residents (Steele et al., 1999) found that crash risk was six times higher after night shifts as compared to day shifts.

Researchers in England, Sweden, and Australia compared crash risk per kilometre driven by times of the day for crashes that did not involve alcohol. Crash risk peaks in the early morning hours, from 2:00 to 4:00 a.m. Crash risk is lowest during daylight hours (Smiley, 2015).

Our driver worked daytime split shifts, with changing start and end times. If these shifts had been at night, they would have resulted in circadian desynchronization. However, the driver's split shifts occurred during daytime hours when he would have normally been awake.

I concluded that, based on the times of day the driver typically worked and slept, and his shift schedule, a fatigue risk related to circadian rhythm desynchronization did not exist.

Type of Work-Rest Schedule

The arrangement of working hours affects the likelihood of short-term and long-term sleep disruption, continuous wakefulness, and circadian rhythm effects, all of which were considered in my fatigue analysis. The driver was working a split shift schedule. This type of schedule involves two distinct working periods per day (Dembe et al., 2006). Our driver would often come home and take naps between shifts. A very large study looked at the relative risk of suffering a work-related injury among people who were working night, evening, rotating, split, and irregular shifts, compared to the risks for workers who were working day shifts. This study found that the risks for split shifts, such as the type our driver worked, were no different from those of day shifts.

I concluded that a fatigue risk related to the type of work schedule the driver worked did not exist.

Sleep Disorders

About one quarter of the population has a sleep disorder; that is, problems getting to sleep or staying asleep. One of the most common sleep disorders is obstructive sleep apnea, which results in short periods of time when breathing stops during sleep. This results in disturbed sleep and, as a consequence, sleepiness during the day. The condition affects up to 1 billion people worldwide (Benjafield et al., 2019).

Our driver was not known to have any sleep disorders nor signs of one (e.g. a body mass index out of normal range, recent weight gain, frequent snoring).

I concluded that a risk of fatigue related to sleep disorders did not exist.

Medical and Psychological Conditions, Illness and Drugs

Our driver had had a cold and sore throat in the week prior to the accident and had called in sick on Sunday, September 14th. The crash occurred on September 18th. He was using over-the-counter throat lozenges and not taking any cold medication. Our driver had type 2 diabetes. However, it was under control through diet and medication.

I concluded that a risk of fatigue related to medical and psychological conditions, illnesses, and drugs did not exist.

Table 8.1 summarizes the findings in relation to fatigue.

Transportation Safety Board Analysis

After a multi-faceted examination of fatigue factors, it seemed unlikely that fatigue was a contributing cause to this crash. So, what were the reasons it occurred? The human factors analysis done by the TSB provides some answers.

At the time of the crash, the bus driver had picked up passengers and was leaving the transit station. The TSB analysis indicated that a key visual task for the driver was that of checking for standing passengers on the upper

TABLE 8.1

Summary of Findings and Conclusions Related to Fatigue

Fatigue Factors	Finding	Conclusion
Recent sleep disruption	The driver had 2 out of 3 prior days with short sleep. There is minor concern about sleep quality.	Fatigue risk may have existed.
Long term sleep disruption or sleep debt	The driver sleep debt at the time of collision was 1.3 hours. At 8 hours, a sleep debt may be considered to be a contributing factor. As long as the sleep debt is greater than 0 hours, fatigue could be a contributing factor.	Fatigue risk may have existed.
Duration of continuous wakefulness	The driver had been continuously awake for 3.7 hours at the time of the crash, considerably less than the 22-hour threshold when micro-sleeps begin.	Continuous wakefulness was only for 3.7 hours. Fatigue risk did not exist.
Time of day or circadian rhythm effects	Driver worked split shifts, with changing start and end times. Split shifts happened during daytime hours when he would have normally been awake.	Fatigue risk related to circadian rhythm desynchronization did not exist.
Work-rest schedule	A study found that the risks of split shifts that the driver worked are no different from those of day shifts.	Fatigue risk related to work schedule did not exist.
Sleep disorders	The driver had no recorded signs of sleep disorders (e.g. a body mass index out of normal range, recent weight gain, frequent snoring).	Fatigue risk related to sleep disorders did not exist.
Medical and psychological conditions, illness, drugs	The driver had called in sick 4 days prior to the crash but was using over the counter medication only. He had type 2 diabetes, which was under control through diet and medication.	Fatigue risk related to medical and psychological conditions, illnesses, and drugs did not exist.

deck of the bus using a video monitor. Typically at station stops, drivers were required to check the in-vehicle video monitor before leaving to ensure that passengers on the upper deck of the double-decker bus were seated. After boarding, some passengers take time to find available seating and end up doing so while the bus is in motion. In this case, the bus driver was required to announce that standing on the upper deck was not permitted.

The monitor provided for the driver to see passengers on the upper deck of the bus had many strikes against it: 1) it was small, 2) it was divided into four screens. It was also poorly configured and inconsistent in the positioning of the four views in the quadrants. It was configured differently depending on the bus, and 3) it was above the driver's line of sight (see Figure 8.4) and to the left (see Figure 8.5). That meant that to use it, the driver had to look up and away from his usual line of sight, down the road. In order to keep an eye on passengers on the upper deck who were still finding seats when the bus left the station, the driver needed to occasionally glance at the monitor while the bus was in motion.

At the same time that the driver might have been looking at the upper-deck passengers, he was also following a left curve as he approached the crossing (see Figure 8.3). As the TSB report says: "Negotiating a curve increases the mental workload of a driver. On a straight road, drivers will tend to look straight ahead and rely on visual stimuli (e.g. telephone poles) passing within the peripheral visual field to guide steering behaviour. When negotiating a curve, a driver must look intermittently toward the vehicle's current and future positions, which are visually separate." Furthermore, the TSB report points out that there were a number of objects (roadway signs, trees, shrubs, horizon line, bus window, pillar) (see Figure 8.1 from TSB Railway Investigation Report R13T0192) that partially obstructed or interfered with the driver's view of the crossing gates and warning lights on the approach.

Between checking the in-vehicle monitor for standing passengers upstairs, conversations about seat availability between the driver and a passenger and among passengers near the driver, as well as the perceived need to make an announcement to passengers standing on the upper deck, a situation existed where the driver was likely distracted in the critical seconds before the

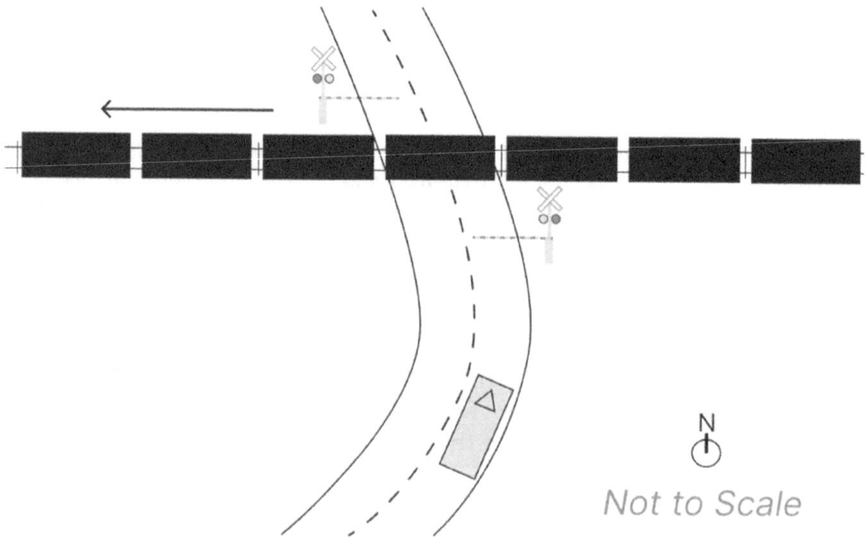

FIGURE 8.3
Overhead view of bus approaching the railway crossing.

FIGURE 8.4
Bus driver workstation.

accident. This would have contributed to an extended perception-reaction time (PRT). PRT refers to the time that elapses between when you detect that "something" is out there, deciding that it is a hazard of some kind and you need to slow down or stop, your brain sending a message to your foot, and the foot to just start to move. It does not include the time it takes you to fully depress the brake or to stop. Typically, in a straightforward situation when there is a hazard on the road, PRT will be on the order of 1–2 seconds.

The driver did not stop the bus before the crossing gates as he was required to. Nor did he slow down until he removed his foot from the accelerator pedal about 3 seconds before impact. As is typically the case when brakes are applied, they only took effect after a short delay, when the bus was approximately 2 seconds from striking the train.

Smiley Fatigue Assessment

In forming an opinion as to whether fatigue was a cause of this crash, many potential sources of fatigue were identified and analyzed. Ultimately, I considered that fatigue was an unlikely contributor for the following reasons:

- The driver had not been awake for very long when the crash occurred.

FIGURE 8.5
View to video monitor.

- The crash happened at a time of day when the driver's circadian rhythm was at a favourable point (when his body was geared for action).
- Despite getting short sleep on 2 out of 3 preceding days, the driver's sleep debt was only 1.3 hours at the time of the crash.
- Based on the driver's work/rest schedule, he was unlikely to have experienced "desynchronised" circadian rhythms.
- The driver had 2 full days off followed by only two out of three subsequent nights before the crash with only a minor concern about the quality of his sleep, suggesting recovery of any sleep debt he might have had.
- Sleep disorders, medical and psychological conditions, illnesses, and drugs likely did not contribute any fatigue risk.

Given my findings, it seemed an open and shut case. It was easy to conclude that it was unlikely that fatigue contributed to this crash. Likewise, on the subject of fatigue, the TSB report similarly concluded that "Based on the review of the driver's work-rest history and related sleep patterns, driver fatigue did not likely play a role in the accident."

TSB Human Factors Assessment

Rather than being an issue of fatigue, the TSB Report concluded that a situation existed where the driver was likely distracted in the seconds before the accident. This would have contributed to an extended PRT. Although the gates had been down for at least 26 seconds, the driver only began to brake when the bus was approximately 3 seconds from striking the train.

For those of you interested in more information about the investigation, I encourage you to go to the TSB website (www.tsb.gc.ca/eng/rapports-reports/rail/2013/r13t0192/r13t0192.html) and obtain the full report – TSB Railway Investigation Report R13T0192.

Countermeasures

Today's vehicles are increasingly outfitted with screens as opposed to knobs and dials. Commercial drivers (police, taxis, maintenance vehicles, transit vehicles, etc.) have specialized displays that provide information related to their work. This was also the case for the bus driver who was the subject of this chapter. As part of our bus driver's job, he had an unnecessarily dangerous task of using a small, poorly configured video monitor positioned high above his head and to the left, while in motion as he departed the station, to verify that passengers were seated in the top deck. That distraction, as well as the related conversations among passengers who were standing near the driver, was found by investigators to have extended his PRT, thereby delaying his response to the train in the crossing, resulting in a tragic outcome. A countermeasure that could address the factors contributing to this bus-train collision would be guidelines and/or regulatory standards concerning the design and location of visual displays used while the vehicle is in motion. In fact, because of this occurrence, the TSB made a formal recommendation (R15-01) to the federal and provincial governments to "develop comprehensive guidelines for the installation and use of in-vehicle video monitor displays to reduce the risk of driver distraction" (TSB, p. 180). The guidelines were published in 2019 and are available at: https://tc.canada.ca/en/road-transportation/stay-safe-when-driving/transport-canada-guidelines-limit-distraction-visual-displays-vehicles.

References

Anund, A., Fors, C., Ihlström, J., & Kecklund, G. (2018). An on-road study of sleepiness in split shifts among city bus drivers. *Accident Analysis & Prevention, 114*, 71–76.

Balkin, T., Thome, D., Sing, H., Thomas, M., Redmond, D., Wesensten, N., Williams, J., Hall, S., & Belenky, G. (2000). *Effects of sleep schedules on commercial motor vehicle driver performance*. Federal Motor Carrier Safety Administration, U.S. Department of Transportation.

Benjafield, A. V., Ayas, N. T., Eastwood, P. R., et al. (2019). Estimation of the global prevalence and burden of obstructive sleep apnoea: a literature-based analysis. *The Lancet Respiratory Medicine*, *7*(8), 687–698.

Coate, D., & Markowitz, S. (2004). The effects of daylight and daylight saving time on U.S. pedestrian fatalities and motor vehicle occupant fatalities. *Accident Analysis & Prevention*, *36*(3), 351–357.

Dembe, A. E., Erickson, J. B., Delbos, R. G., & Banks, S. M. (2006). Nonstandard shift schedules and the risk of job-related injuries. *Scandinavian Journal of Work, Environment & Health*, 32(3), 232–240.

Grandjean, E., (1982). *Fitting the task to the man: An ergonomic approach*. Taylor & Francis Ltd.

O'Neill, T. R., Krueger, G. P., Van Hemel, S. B., & McGowan, A. L. (1999). *Effects of operating practices on commercial driver alertness* (No. FHWA-MC-99-140). Federal Highway Administration.

Pternitis, C. (1977). Etudes electrophysiologiques au cours du sommeil chez l'ouvrier poste. In P. Andlauer and collab. (Eds.), *Ergonomie du travail de nuit et des horaires alternants*. Editions Cujas.

Rudin-Brown, C. M., & Rosberg, A. (2021). Applying principles of fatigue science to accident investigation: Transportation Safety Board of Canada (TSB) fatigue investigation methodology. *Chronobiology International*, *38*(2), 296–300.

Smiley, A. (2015). Fatigue and driving. In A. Smiley (Ed.), *Human factors in traffic safety* (pp. 277–293). Lawyers & Judges Publishing Company.

Steele, M. T., Ma, O. J., Watson, W. A., Thomas, H. A., & Muelleman, R. L. (1999). The occupational risk of motor vehicle collisions for emergency medicine residents. *Academic Emergency Medicine*, *6*, 1050–1053.

The Transportation Safety Board of Canada. (n.d.). Railway Investigation Report R13T0192. Retrieved February 19, 2025, from https://www.tsb.gc.ca/eng/rapports-reports/rail/2013/r13t0192/r13t0192.html

9

Car Versus Tractor

Introduction

A driver in a rural area encountered what she thought was an oncoming vehicle with one headlight moving into her lane. She began pulling over to the right to avoid it. After 10 seconds she rear-ended the vehicle, which was a tractor with a white work light on the back. The scientific issues important in this case were expectancy, pattern recognition, perception of closing speed, and glare.

The Collision

According to the police report, the crash happened at 10:20 p.m. in April on a county road. Ms Sarah Magor was driving a Lincoln. She was accompanied by her friend Ms Campbell. The two women were going home after playing bingo at a local church.

Mr Farmer was driving a Massey Ferguson tractor towing a spraying unit equipped with a work light right next to a black-on-orange "SLOW-MOVING VEHICLE" sign. He had just finished spraying an apple orchard on the west side of the road and was going to head for the orchard to the north. Both vehicles were northbound at the time of the accident. Mr Farmer had pulled out from an orchard laneway onto the county road. According to Ms Magor, as she crested a rise in the road, she saw a white light in the southbound lane. Her car was equipped with an automatic headlight dimming system, which dimmed her high beams as she approached the tractor. She was under the impression that there was an oncoming vehicle with one headlight encroaching into her lane (see Figure 9.1). She hit the tractor, resulting in the death of Mr Farmer.

DOI: 10.1201/9781003564805-9

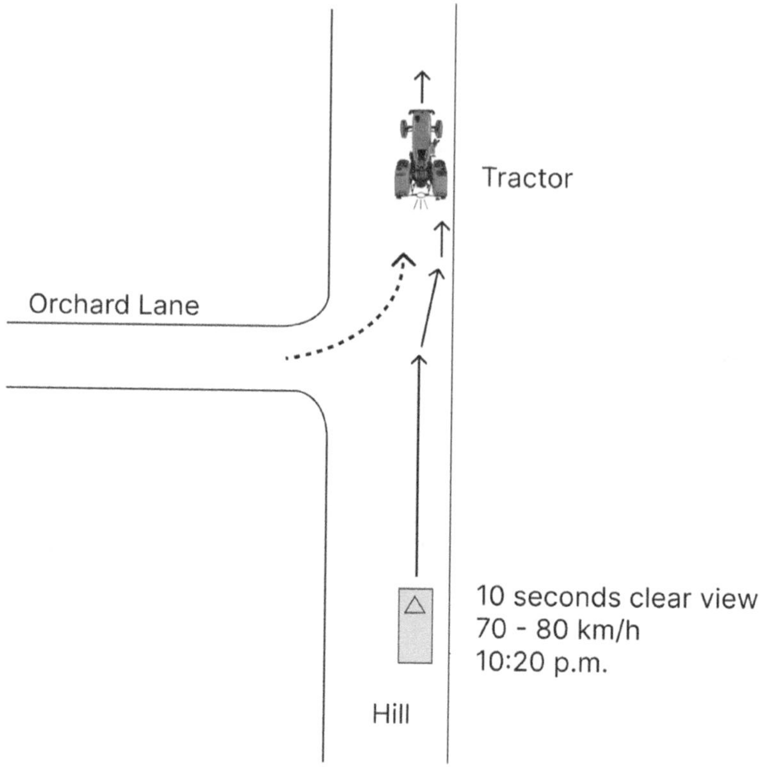

FIGURE 9.1
Paths followed by involved drivers.

The Investigation

I visited the site of the crash accompanied by two collision reconstructionists, Mr Alan Archer and Mr Steve Mitchell of Clinton Engineering. I made observations of a Massey Ferguson Model 194 tractor to which was attached a sprayer and a work light similar to those involved in the collision. According to Mr Archer, the tractor we observed was similar in terms of acceleration capabilities to the tractor involved in the accident.

During the afternoon, I observed the tractor pulling out of an orchard laneway on the north side of the road and pulling across the road into the northbound lane up to the point of impact. The time taken between the tractor entering the roadway from a stopped position and reaching the point of impact was approximately 15 seconds. With the trajectory driven by Mr Mitchell, the work light crossed into the northbound lane within the first 2–4 seconds.

After dark, I again observed the tractor and sprayer. The work light appeared equivalent in brightness to a car low beam. In the first few seconds of the tractor's travel onto the road, the path of the work light from the southbound lane to the northbound lane could be mistaken for that of an oncoming car with one headlight.

In order to determine Mr Farmer's opportunity to avoid this crash, I checked the visibility of northbound traffic for him as he would have seen it from the tractor at the edge of the road. Approaching the site going northbound, there is a steep hill, followed by a level area and a rise. Headlights of northbound cars are first visible by means of the headlight loom and then the headlights themselves become visible.

There was very little traffic in this rural area. The vast majority of vehicles we observed were using high beams. Therefore I first observed Mr Archer's car which he drove northbound at 80 km/h (50 mph) with the headlights on high beams. The headlight loom was visible about 30 seconds before the point of impact. The headlights were in full view for about 20 seconds before reaching the point of impact, with the exception of a 4 second period when the headlights disappeared in a dip but the loom was still clearly visible. For the same speed, but on low beams, the loom was first visible about 19 seconds before the car reached the point of impact. In other words, the oncoming Lincoln could be seen by Mr Farmer had he looked to the north for about 16 seconds as it approached.

The Science

The scientific issues important in this case were expectancy and pattern recognition, perception of closing speed, and the effect of glare on visibility distance.

Expectancy and Pattern Recognition

Based on my observations at the site, I concluded that a driver viewing a white work light crossing from one lane into the other would easily be misled into thinking that he or she was approaching an oncoming vehicle. At night visual information is greatly reduced and drivers rely on previous experience to interpret what they see. Oncoming vehicles are equipped with white lights at the front and red lights on the back. The white light on the tractor was equivalent in brightness to a low beam headlight. It would be a natural assumption for a driver seeing such a white light on the road at night to think it belonged to an oncoming vehicle. This impression would be supported by the fact that the Lincoln headlights dipped in response to the appearance of this light, as for oncoming cars. The impression would be further strengthened by observing that the gap between the vehicle ahead and Ms Magor's own car was closing, as would be the case for an oncoming vehicle (more on this later).

Lights and reflectors do not guarantee longer seeing distances to objects in the road at night. If the viewer does not know to what they are attached, they may not be able to use the information to "connect the dots" and recognize the pattern. This was shown in the study of pedestrian recognition where pedestrians were equipped with retroreflective discs. Unless the drivers knew that the discs indicated were carried by pedestrians, the visibility distance was no longer for pedestrians so equipped as compared to pedestrians not equipped (Shinar, 1985).

The tractor was in view to Ms Magor for over 10 seconds, a long distance when driving at high speed. She was much criticized for not figuring out, with that length of time, what was going on. However, drivers vary greatly in how long it takes to recognize a pattern, in this case to determine that the white light belonged to a vehicle moving away from the Lincoln and not toward it and it belonged to a slow-moving tractor. In a course on human factors in reconstruction, given numerous times to police, I illustrate the difficulty of pattern recognition by presenting an incomplete image, and timing how long it takes for viewers, given either no information or some information (e.g. it is an animal) about what is in the picture, to recognize the pattern present. The picture is of a dog made of black blobs on a white background. There is wide variability. Some viewers recognize the pattern as a dog within a few seconds. On average, those who are given some information about what is in the picture respond much more rapidly with the name of the breed of dog (Dalmatian) and those given no information. However, whether or not viewers have been given information about what is in the picture, the variability in time to recognize a pattern in the picture is considerable, some recognizing it within a few seconds, and others who still cannot make out the pattern after receiving information about it and having had several minutes to look at it.

Perception of Closing Speed

One might argue that Ms Magor ought to have been able to tell the difference between closing on a slow-moving vehicle with a white light on the rear and closing on an oncoming vehicle. However, drivers are very poor at making such differentiations. At night the main cue that one is approaching another vehicle is the changing separation of the taillights or headlights as the case may be. Even when there are two lights, creating a wide angle at the eye, drivers' estimates of the speed of oncoming traffic are very inaccurate. In fact, in an overtaking situation drivers cannot distinguish between cars approaching at faster than the speed limit and drivers approaching below the speed limit until they are very close.

In the case of the collision Ms Magor had only **one light** from which to judge the rate of closure. The main cues to the fact that she was approaching the light would come from the increase in the apparent size and brightness of the light. These cues are even more difficult to use to estimate the approach speed than is the cue of changing separation of the lights. As Ms Magor approached the tractor light, she would be able to detect that the gap between the light and herself was decreasing. Detecting the gap was not a problem. The problem was that it would be very difficult for her to accurately judge how **quickly** the gap was closing and to realize that she was gaining on a slow-moving vehicle as opposed to approaching an oncoming vehicle.

How quickly does the gap have to be closing for a driver to recognize that a rear-end collision or head-on collision is imminent? Laboratory studies have been used to assess drivers' ability to judge closing speed and distance to an approaching vehicle (whether the driver is approaching a slower vehicle ahead or an oncoming vehicle). It has been determined that the most important cue is the rate of change of the apparent size of the approaching vehicle at the eye of the observer (Dewar & Olson, 2015 in Smiley, 2015, p. 30). At a distance, the apparent size of the approaching vehicle is very small. As the distance closes, the apparent size of the vehicle increases very slowly at first, and then more and more rapidly. For those of you who remember high school mathematics, this is a tangent function.

Research carried out by Hoffmann and Mortimer (1996) found participants in laboratory experiments do not **begin to perceive** that they are closing on a lead vehicle until the apparent size of the approaching vehicle expands at a rate of 0.2°/second. Furthermore they do not perceive the lead vehicle to be an "immediate" hazard until the apparent size of the approaching vehicle expands at just over twice the rate at which it can first be detected, or at 0.4°/second (Muttart et al., 2005). The distance at which drivers can first start to recognize that they are closing rapidly can be calculated as shown in the text box. It is 31 m (102 feet) which gives the driver, travelling at 80 km/h (50 mph) or 19.4 m/s (64 ft/s), only 1.5 seconds to respond. Barely enough time for perception-reaction but not enough time to swerve or stop.

For those who would like details on how to calculate the distance at which drivers become aware that the gap is closing quickly please read the text box below entitled "A Little Math". Do not be scared.

A LITTLE MATH

The distance ($D_{(th)}$) from the vehicle in front, in this case the tractor, at which the rate of change of the width of the work light starts to exceed the bare minimum of 0.2°/second or 0.003 rad/second (1 rad equals 57.3°) (T) is determined by:

$$D_{(th)} = \sqrt{\frac{W \times T}{V}}$$

In this situation, let us say that the tractor was going at 10 km/h, while Ms Magor, who was very familiar with these roads was travelling at the speed limit of 80 km/h. Thus, the closing speed (V) was 80 km/h minus 10 km/h or 70 km/h which is equivalent to 19.4 m/second. Assuming a work light width of 15 cm (0.15 m), a little calculation with the equation above will show that the distance at which a light of that size, which is all that would be visible, would begin to be detected as closing rapidly would be 31 m. (T is threshold distance.) This gives the driver 1 ½ seconds to respond. This is enough time for perception-reaction but not enough time to swerve or stop.

Glare reduces seeing distance because it effectively reduces the contrast of the object being viewed against its background. The closer the glare source is to the driver's line of sight, the greater its impact on visibility. The work light was right next to an orange and black "SLOW-MOVING VEHICLE" sign. The white light would have been a source of glare making it very difficult to see either the tractor or the warning sign attached to it.

Age and Glare

As drivers age, their sensitivity to glare increases, and they take longer to recover after being exposed to it. In the situation of the crash, the white work light next to the "SLOW-MOVING VEHICLE" sign had been present for at least 10 seconds. Based on our observations of the crash site, the work light would have been more disabling than a passing set of headlights. Glare from the work light would have made it very difficult for Ms Magor to detect the tractor and the "SLOW-MOVING VEHICLE" sign. Both of these targets were dark in colour and not well contrasted against the road or each other.

At the time of the crash Ms Magor was 70 years old. A study of older drivers and glare required subjects to continuously adjust the brightness of a target disc until they could just see it starting with when a glare source had been turned off. It took the older subjects twice as long to recover their ability to detect the target. Furthermore, for older subjects the target disc had to be about twice as bright as for the younger subjects at the ambient level to be visible (Olson & Sivak, 1981). Thus, in the presence of glare, older drivers are at a significant disadvantage compared to younger drivers.

The Analysis

When I viewed the trajectory driven by Mr Mitchell at the collision scene, he crossed the centre line within 2–4 seconds and continued down the road for another 11–13 seconds to reach the point of impact. However, what Ms Magor experienced would depend on whether Mr Farmer followed the same path as Mr Mitchell, or whether he crossed the road at a more oblique angle, extending the time he spent in the lane for oncoming traffic. In any case, he would have ended up centred in the northbound lane at the time of impact. A white light positioned at the centre of the lane could easily be misinterpreted as belonging to an oncoming vehicle with a missing headlight halfway into the lane, especially if one had also viewed the light moving from the southbound to the northbound lane as Ms Magor had (see Figure 9.1).

Ms Magor continued to believe she was facing an oncoming vehicle for a period on the order of 10 seconds or more. Once one is locked into one way of perceiving a scene it can be difficult to change, especially without obvious evidence to the contrary. It is a characteristic of human decision making that we look for ways of confirming our interpretation rather than for disconfirming evidence (Tijerina, 2015). Ms Magor states that she was not able to make out the vehicle on which the light was mounted, nor the "SLOW-MOVING VEHICLE" sign. This is to be expected given the glare from the light and the low contrast of the dark green tractor and sprayer against the road. Had she been able to see the outline of the tractor, Ms Magor's perception of what was happening would likely have changed.

Upon seeing what she thought was an oncoming vehicle moving into her lane, Ms Magor lifted her foot off the accelerator and started to brake, evidently not hard enough to leave skid marks or slow the car appreciably. She moved over in an attempt to avoid the "oncoming vehicle" which appeared to be "hogging the lane".

In such a situation, an avoidance manoeuvre of pulling over to the side of the road is more appropriate than an emergency braking response. The latter would not avoid a head-on collision. The former would have a greater chance of success. Ms Magor moved over in her lane sufficiently that she would have been very close to avoiding contact, had the tractor light really been a low beam headlight on the driver side of an oncoming car, instead of a light centred on the 1.2 m wide sprayer.

Conclusions

Ms Magor was confronted with very misleading visual cues. At night, because of the reduction of visual information, drivers must rely on experience in interpreting what they see. Ms Magor had three strong cues telling her she was facing an oncoming vehicle:

1) The light she saw was white, with equivalent brightness to a low beam headlight found on oncoming vehicles.
2) Her headlight beams dipped in response to its appearance, as they did for lights of oncoming vehicles.
3) The gap between her vehicle and the white light closed, as it would for an oncoming vehicle.

While drivers easily can tell that the gap is closing, the fact that there was only one light made it exceptionally difficult to estimate the rate of closure of the gap and realize that she was gaining on a slow-moving vehicle rather than approaching an oncoming vehicle. The brightness of the work light and the low contrast of the sprayer and tractor against the road prevented her from seeing all but the work light on the vehicle she was approaching. Even after hitting the sprayer and tractor, she thought she had hit a four-wheeled ATV.

In my opinion, the average reasonably attentive driver would be just as likely as Ms Magor to misperceive the scene ahead. Based on my knowledge of human perceptual and information processing processes (expectancy and pattern recognition), her misperception of the situation is entirely understandable.

Countermeasures

Retroreflective striping has been found to be very useful at night in helping drivers to understand the nature of and movements of (e.g. backing up across a highway) vehicles ahead of them. In 2001, the U.S. Federal Motor Carrier Safety Administration (FMCSA) required the application of retroreflective striping on all heavy truck trailers. In 2001, a report was published on the effectiveness of the striping in enhancing the visibility of heavy trailers and reducing side and rear impacts by other vehicles into these trailers during dark conditions. It was based on a statistical analysis of almost 11,000 crash cases. The tape was found to be quite effective. Striping was especially effective in reducing injury crashes. In dark conditions, it reduced side and rear impacts that resulted in fatalities or injuries to drivers of any vehicle by 44% (Morgan, 2001).

The FMCSA report indicates that retroreflective striping outlining the tractor would have assisted in making the unusual outline of the tractor visible and might have prevented this crash. The same could be applied to other farm equipment that uses public roads.

References

Hoffmann, E. R., & Mortimer, R. G. (1996). Scaling of relative velocity between vehicles. *Accident Analysis & Prevention, 28*(4), 415–421.

Muttart, J. W., Messerschmidt, W. F., & Gillen, L. G. (2005). Relationship between relative velocity detection and driver response times in vehicle following situations. *Society of Automotive Engineers Technical Paper 2005-01-0427*, SAE 2005 World Congress, Detroit, Michigan, April 11–14, 2005.

Morgan, C. (2001). *The effectiveness of retroreflective tape on heavy trailers* (NHTSA Technical Report No. DOT HS 809 222). U.S. Department of Transportation.

Sivak, M., Olson, P. L., & Pastalan, L. A. (1981). Effect of driver's age on nighttime legibility of highway signs. *Human Factors, 23*(1), 59–64.

Shinar, D. (1985). The effects of expectancy, clothing reflectance, and detection criterion on nighttime pedestrian conspicuity. *Human Factors, 27*, 327–333.

Tijerina, L. (2015). Information processing, attention, and driver workload. In A. Smiley (Ed.), *Human factors in traffic safety* (3rd ed, p. 39.). Lawyers & Judges Publishing Inc.

10

Work Zone Crash

Introduction

A young man was driving two friends to their homes late one rainy night. According to police, he was going too fast when he entered a curve. His vehicle left the road and struck a large rock.

The Collision

The crash happened early in the morning in the spring on Highway 1, just north of the junction with County Road 12 in a 60 km/h speed zone. It was dark and raining at the time. According to his statement to the police, Mr Ted Lennon, an eyewitness to the crash, had just come over the crest of the hill on Windsor Road driving southbound when he saw a flash and his passenger heard the crash.

Figure 10.1 shows an aerial view of the split between Highway 1 and Windsor Road at the distance from the split where Windsor Road appears to be the extension of Highway 1. The driver's line of sight to the split in Highway 1 between Windsor Road, the former continuation of Highway 1, is shown in Figure 10.1 and Figure 10.2. Figure 10.1 on the left shows the driver's eye view of the split. Figure 10.2 on the right shows the aerial view of the split.

The new portion of Highway 1 lies along the right curve. The dashed line in Figure 10.2 shows the northbound driver's line of sight. Mr Young (aged 19) was driving a passenger vehicle north on Highway 1 accompanied by two male friends. The police report describes Mr Young as veering left onto the shoulder and continuing into a large rock, at the junction of the new Highway 1 (which curved to the right) and Windsor Road (the former straight continuation of Highway 1). As a result of the impact, the vehicle burst into flames, resulting in the deaths of Mr Young and his two passengers. According to his statement to the police, Mr Ted Lennon, an eyewitness

DOI: 10.1201/9781003564805-10

FIGURE 10.1 AND FIGURE 10.2
Showing the main road (reconstructed Highway 1) curving to the right and the straight road portion that could be perceived as the continuation of Highway 1. Dashed line shows driver line of sight.

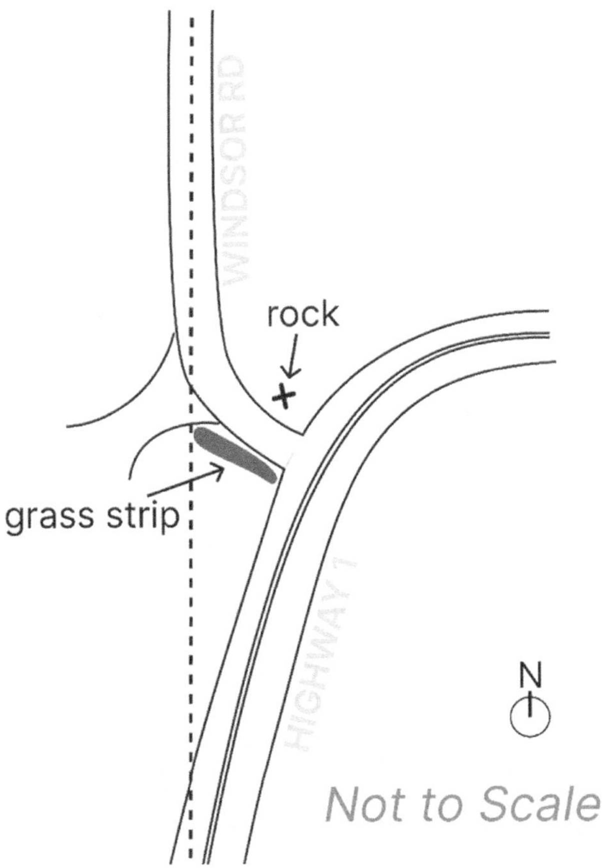

FIGURE 10.2
Continued

to the crash, had just come over the crest of the hill on Windsor Road driving southbound when he saw a flash and his passenger heard the crash.

The Mandate

For me, the human factors question was: given the circumstances, would most reasonably attentive but unfamiliar drivers have realized that the new road curved to the right? Would they have realized it in time to follow the curve rather than going straight? Or was that not clear from the road path, signing and marking?

The Investigation

Sometimes it is possible, though not advisable, to carry out an investigation without going to the site, or without going to the site at the same time of day and under the same weather and road conditions. This is especially true if one is hired months or years after a collision involving a site which has been changed since the time of the crash. In this case, I was asked to visit the site a mere 19 days after the collision. Only in cases involving transient situations such as work zones or just prior to some critical aspect of the road being changed, do I see a site so soon after a collision.

When visiting the site, it is important not just to consider the layout and traffic control devices at that site, but also the traffic conditions, the lighting, and the weather conditions at the time of the crash. The crash had happened at 2:00 a.m., when it was fully dark. For reasons of convenience, I chose to arrive at the site at 10:00 p.m. This is not a concern because at this time, it was as fully dark as it would be at 2:00 a.m., and light levels would be very similar to those at the time of the crash, as would traffic levels.

The crash had happened when it was dark and raining. One always wants to visit the site under the same circumstances as the crash occurred. It's easy enough to arrange for a visit in darkness, but rain is another matter! I remember researchers in the United Kingdom talking about trying to study visibility in fog, a similar challenge. They would get notified when a crash had occurred in fog, tear off to the site to measure visibility distance, and invariably find that the fog had disappeared by the time they got there. They decided that the main finding from their research was that the best way to get rid of fog is to study it!

The effects of rain are a little easier to investigate. When a collision involves rain, I pick times for the site visit when the forecast is for a high chance of rain. However, the weather does not always cooperate, especially if it takes a long time to reach the site. On this occasion, my timing for visiting the site was fortuitous. When I left home it was dark and raining, and the rain continued for most of the site visit. I noted that the pavement markings became very difficult to see under these circumstances. It is always reassuring to the court if an expert can say they have experienced the same situation as the person involved in the collision.

I approached the site as Mr Young had, along a highway through a rural area, interrupted by a few intersections with buildings and the occasional streetlight. The approach to the crash site was through a work zone.

The distance from the start of the work zone up to the crash site, north of the intersection, was approximately 0.6 km (0.4 miles). In this short distance, there were 15 retroreflective traffic signs on the right-hand side of the road. The seventh of these signs was a warning sign indicating that the main road, Highway 1, curved (see Figure 10.3).

The row of streetlights on Highway 1 lined up with the streetlights on Windsor Road, making it appear as if Highway 1 continued straight along the old alignment (see Figure 10.3)

The "MAIN ROAD CURVES" sign is circled on the right. There are no work zone signs in the photograph, but there were at the time of the crash and of my site visit. At the time I saw the work zone, 19 days after the crash, the warning sign that the road curved was immediately next to a lower sign showing a bump on the road and just prior to an overhead flashing light sign at the intersection. Whether these signs were in the same location on the night of the crash is not known, and it is not possible to tell from photographs taken on the night of the crash. The visual impression of the work

FIGURE 10.3
Daytime view approaching crash site. (Photo taken months after the crash, and after the work zone had been cleared away.)

zone from the photographs, and on the night of my site visit, was that it was cluttered with many reflective objects, among which was a sign with the critical information that the main road curved.

As I approached the curve in Highway 1, there were no strong indications that the road was curving, other than the white edge line. The centre line marking was very worn and not very visible in the rain.

The only break in the straight-ahead path was a grassy strip running across the former continuation of the highway, between the first and second streetlight. The grassy area contrasted with the road surface. However, this area did not appear to be very wide, because of the slight banking of Highway 1, and could easily be interpreted as a dark area of pavement between the two streetlights on a road that continued straight ahead (see Figure 10.1).

Altogether, the site visit was most revealing about why Mr Young found himself initially going down the wrong path.

Investigator Information

An investigator is often hired to take photos or dig up information, such as sleep schedules or drinking habits, of individuals involved in a crash. In this case, along with other things, an investigator made inquiries of Mr Young's friends as to how familiar Mr Young would likely have been with the section of Highway 1 in which the crash occurred. It seems that Mr Young had likely been on the road on a few occasions but would not have been very familiar with it.

The Science

After considering my observations made at the site visit, and Mr Young's likely unfamiliarity with the site, I decided that the scientific issues which were important in this case concerned driver information processing load, driver expectations, preview of the road ahead necessary to steer accurately around a curve, and PRT.

Driver Information Processing Load

Based on my observations at the site and on the photographs taken the night of the crash, the visual impression of the work zone that preceded the split in the road would likely have been that it was cluttered with many reflective objects.

Drivers' capacity to process information is limited. "Information" that we "process" refers to matters such as the colour of a sign, the placement of our vehicle, the words of our passenger, etc. As we travel through space, we are confronted with masses of information, only a small part of which we actually notice or process and retain, even briefly. Information has been measured in bits per second. A bit of information is equivalent to the answer to a single yes/no question. It has been estimated that, of over 1 billion bits per second of information directed at our sensory systems, we are consciously aware of only roughly 16 bits per second of new information. In other words, we could answer about 16 yes/no questions related to a new scene that we had not experienced before and saw the scene for one second (Grandjean, 1988).

Human capacities were designed for walking and running speeds – not for the speeds that drivers travel on the highway where they move through space at 20 m/s (75 ft/s) or more. The faster we move, the more we are taxing our information processing capacity. It is for this reason that roads are designed in a very consistent manner and that the faster the speed limit, the simpler the design. For example, freeways have the simplest design: opposing traffic is divided, traffic goes in one direction, grades and curves are very gentle, there is no requirement to respond to traffic lights or stop signs, and there are no pedestrians, bicyclists or parked vehicles. At freeway speeds, drivers would not be able to process information from these sources rapidly enough to avoid crashes. When such features are present, speed limits are considerably lower, reflecting the need to slow the rate of information with which drivers must contend.

To compensate for the limited processing capacity of drivers, information presented to drivers should be managed carefully to avoid overloading them. The advice given in the *Institute of Traffic Engineers (ITE) Manual*, Chapter 3 'Road Users' is: "Where all the information required by the driver cannot be placed on one sign or on a number of signs at one location, spread it out along the road so that information is given in small chunks, thereby reducing the information load on the driver" (Smiley & Dewar, 2016).

Assuming the same type of work zone signing was in place on the night of the crash as during my site visit 19 days later, the critical sign that announced that Highway 1 curved appeared amongst a clutter of other signs. The critical sign was placed prior to the Highway 1 split into the old straight continuation of Highway 1 (now Windsor Road) and the newly reconstructed section of Highway 1 that curves to the right.

A second concern was that this critical curve warning sign was also immediately adjacent to a lower sign, competing for the driver's attention, showing a bump on the road, and just in advance of an overhead flashing light sign, competing even more strongly for the driver's attention, at the intersection.

Unfortunately, the critical information that the "main road curves ahead" was followed with conflicting sign information that the main road goes straight. The photo in Figure 10.3 was taken by an investigator 4 months after the crash when the work zone was long gone. It shows that the sign warning that the main road curves, in advance of the intersection, was followed by a street name sign indicating that Windsor Road continues straight ahead.

At this point the straight path ahead would be in full view. The Highway 1 curve to the right would be obstructed from view by the buildings and foliage.

Thus the critical sign that the road curves was:

- buried in a clutter of potentially distracting other work zone signs and an overhead flasher – some of which were very attention-getting
- contradicted by a subsequent sign indicating that the road went straight
- located in a position such that the curve to which it referred was obstructed from drivers' view

Although he was going too fast for the curve, as the police said, Mr Young's expectation was that the road continued straight and a higher speed than the speed limit would not be a problem.

Driver Expectations

Drivers' expectations about the path of the roadway are formed by a number of cues. Expectation is key because it is a powerful determinant of accuracy and reaction time. To take a very simple example, as you enter a dark room that you have never entered before, you expect the light switch to be about shoulder height and near the edge of the door. In North America you also expect to move it up to turn it on. If it is placed as expected, and operates as expected, you can locate it and turn the light on quickly. If it is not placed as expected – say at waist height, and operates differently than expected, for example, it moves laterally instead of vertically, your response time will be considerably longer. Similarly, drivers will respond quickly and accurately when roads and traffic control devices are designed according to their expectations. "Design according to driver expectations" is a principle that has been taught to traffic engineers and highway designers for decades.

When drivers are surprised because their expectations are violated, slowed responses and errors occur. An example of such a violation of expectations occurs when a freeway exit is located on the left side rather than the right side. It has been known since the 1960s that left exits are associated with over twice the crash rate of right exits. Another example of violation of expectations is when street name signs are located on the left side of the road in left hand drive countries. A study of older drivers using street name signs to follow a route showed drivers made significantly more mistakes when signs were placed on the left side than when on the right (Smiley & Smahel, 2011). In another situation, signs were placed much lower down than usual, on property fences. These signs were also missed by drivers navigating in unfamiliar areas and in environments considerably less complex than the one facing Mr Young.

Mr Young's expectations as to the path he should follow were likely violated by the situation that he encountered. The effect of expectations is described in detail in *A User's Guide to Positive Guidance* (Post et al., 1981), a psychology-based approach to highway design, which was developed by the U.S. FHWA in the 1970s. The authors discuss driver expectations in a situation very similar to that encountered by Mr Young, in which the main road split into a curved portion and a straight (tangent) portion which was the old continuation of the road. The authors state that "a line of utility poles, trees or railroad tracks running parallel and adjacent a long section of tangent roadway creates an expectation that the condition continues on indefinitely... When the highway then curves away from poles or a tangential (straight roadway begins at the point of curve), a hazardous condition is created for the driver" and "run-off-the-road accidents are likely to occur. The function of Positive Guidance at such a location is to change driver expectation by obstructing the visual line of sight or by the use of strategically placed warning signs or delineators showing alignment."

Figure 10.1 was taken south of the area of impact. If you imagine this scene at night, you can see how a driver might be misled into thinking that Highway 1 continued straight through. At night in the spring (without snow as in Figure 10.1), in the rain, that strip of grass, separating Highway 1 from Windsor Road, would appear dark. It could easily be mistaken for an asphalt patch as opposed to a divider between two roadways.

But that is not the end of the challenges! Just at the very moment Mr Young was approaching the split in the roadway, an oncoming vehicle with its headlights lit descended the hill on Windsor Road. Imagine the effect that would have on your impression of where the road went and therefore where you should steer your vehicle.

As an aside, I have said that "Mr Young's expectations as to the path he should follow were probably violated by the situation that he encountered." This is less certain than "in all probability" or " likely" and you have to be very careful with those words. I once gave conflicting evidence about the sharpness of a curve. In one place in my report I characterized the curve as "sharp" and in another I referred to it as "very sharp". That lead to intense questioning as to what I really meant!

Preview Required to Negotiate a Curve

A certain amount of preview is required by drivers to accurately follow a curve. We rely on our vehicle headlights, roadside signs and pavement markings to guide us along the roadway, at night especially. Edge line and centre line marking allows drivers to determine their position within the lane. It also provides warning of changes in road direction. In order to stay in the lane, studies of tracking accuracy suggest that a minimum of 2 seconds' preview is necessary (McLean & Hoffman, 1973). Two seconds' preview means that, if you are travelling at 80 km/h (50 mph), or 22 m/s (72 ft/s), then 2 seconds' preview is 22 m/s (72 ft/s) multiplied by 2, or 44 m (144 ft). In other words, if you were to travel at 80 km/h, to do so comfortably, you would need to be able to see at least 44 m ahead. To determine changes in road direction, even more time, 3–5 secs, is required. Thus, drivers are most likely to be looking a minimum of 3 seconds ahead.

Visibility of Lane Markings

It was raining at the time of the crash and the roads were wet. Wet pavement reflects less light than dry pavement and so appears blacker. Rain makes it difficult to see lane markings, as I observed on my way to the crash site. The newer the lane markings were, the more visibility they would offer. This is because lane markings are made up of retroreflective beads, which in dry conditions collect the light from the headlights and send it back along the same angle as the light struck the pavement, back to the headlights, near the driver's eyes. However, rain fills in the spaces between the beads, creating a smooth surface, which reflects light equally in all directions, greatly reducing the light that reaches the driver's eyes, and making the pavement markings considerably more difficult to see.

Now, the situation here with respect to the visibility or preview distance offered by the lane markings was complicated. We did not know whether 4 months after the crash, when Figure 10.1 was taken, the lane markings had been newly painted or whether they had been present at the time of the crash. At the time, I was not aware of any experiments that had tested the visibility of worn and wet lane markings under streetlights. In such a situation, it is the privilege of an expert to be able to declare "in my opinion" and to state what you think scientific studies, were they to be done, would show. It is for this privilege that an expert thoroughly familiarizes him or herself with literature on the subject and undergoes a lengthy qualification process before being allowed to give opinion evidence. My opinion was that the preview distance of the markings would have been inadequate. This opinion was based on the following information.

A study of the visibility of newly painted lane markings using new headlights on low beams in clear weather conditions found they could be seen on average at 73 m (240 ft) from an observer. In contrast, old markings in a wet condition with used headlights were visible at less than half the distance (Blaauw & Padmos, 1982).

Forensic engineers estimated that Mr Young was travelling at a speed of 70–80 km/h (43–49 mph) as he approached the mid-point of the curve. In this speed range, he would have been able to see new, wet markings approximately 3.5 seconds ahead of him and old, wet markings at approximately 1 second or less. I concluded that this is insufficient preview to provide adequate warning of changes in road direction based on lane marking alone.

Perception-Reaction Time (PRT)

According to the engineering investigation, as Mr Young's vehicle crossed over the grassed area it started to rotate in a clockwise direction, suggesting the driver may have attempted to turn right to try and return to the main roadway, before striking the large rock. In other words, he realized his error, and initiated a response, but it was too late.

PRT refers to the time that elapses between when you detect that "something" is out there or is happening, deciding that it is a hazard of some kind, and that you need to do something (e.g. slow down or stop), your brain sending a message to your foot, and the foot to just start to move. It does not include the time it takes you to fully depress the brake pedal or to stop the vehicle.

Typically, in a straightforward situation when there is a hazard on the road, driver PRT will be on the order of 1–2 secs (Olson et al., 1984). However, the situation facing Mr Young was not nearly as simple to interpret and his response would likely have been at the longer end of the PRT range. Because of the streetlights on Windsor Road, he would have had a view of the straight path ahead for about 4 seconds before reaching the critical split in the roadway. His view of the curving main road to the right would have been blocked by buildings and trees. The physical evidence of his attempt to turn right to return to the main roadway suggests that Mr Young realized his error on the order of 2 seconds before his vehicle struck the rock, enough time for PRT, but not enough to safely respond.

Use of Photos in Court

The two photos in this chapter (Figure 10.1 and Figure 10.3) were not used in court for this case. Had they been, the differences between the view they presented and that which was encountered by Mr Young just prior to the crash would have to be carefully outlined. Otherwise, the photos could be very misleading.

The photos are useful for giving a general impression of the site but they differ from the actual scene at the time of the crash in a number of critical ways. In the photos it is daytime, and the roads were dry. Preview distance, that is, the distance the driver could see ahead, would be much further during the day than at night. In the winter, the thin strip of grass that separated Highway 1 and Windsor Road was covered with snow and would have made the separation of the new road and the continuation of the old road more obvious than it would have been at the time of the crash. As such, this photograph understates the difficulty of determining the road path. The critical warning sign that the road curves ahead would have been seen in the presence of the many signs that were there on the night of the crash, due to the work zone.

Conclusions

A number of human factors limitations were important in this case. First, there was a high information processing load due to the high number of signs and the placement of the critical curve warning sign approaching the critical split. Second, there were a number of cues which could easily have briefly led Mr Young to the conclusion that, at the split in the roadway, Highway 1 continued straight. There was a lack of visual separation between Windsor Road (the former continuation of Highway 1) and the new portion of Highway 1 that curved to the right, making it appear that Highway 1 continued straight. The location of the streetlights along the old section of Highway 1 would have added to this illusion. Third, the lane markings did not provide enough visibility to guide the driver along the highway. Fourth and finally, in the seconds before the crash, a vehicle descended the hill on Windsor Road. It would have looked like opposing traffic on Highway 1. Up to the split in the roadway, Windsor Road would have looked like the former continuation of Highway 1. Is the appearance of this vehicle what caused Mr Young to briefly think that the road continued straight? Although the engineering analysis indicated that Mr Young appeared to have started to turn back towards the correct lane while he was still on Highway 1, it was too late to safely manoeuvre the vehicle around the curve.

In my opinion, the combination of these circumstances created a trap for an unfamiliar driver, and misled Mr Young briefly, but for a critical few seconds, into thinking that the road continued straight. Although Mr Young appeared to have started to perceive his error while he was still on the road, it was too late to safely manoeuvre the vehicle around the curve. The presence of a large rock so close to the roadway was an unnecessary hazard which turned what in most cases would have been a minor crash into a tragic one.

Other factors could have contributed to Mr Young's error – distraction for example, given the young age of the male passengers, or alcohol. While participation in distracting activities is a common practice, leading to increased crash risk over that of non-distracted driving, there is no evidence one way or another that Mr Young was so engaged. Forensic analysis ruled out alcohol as a factor.

Countermeasures

As discussed above, Mr Young's expectations as to the path he should follow were likely violated by the situation that he encountered. The effect of violated expectations is described in detail in *A User's Guide to Positive Guidance* (Post et al., 1981). With strong cues (especially the line of streetlights) pointing towards following the straight portion of the roadway, and an overwhelming number of signs in a short distance within which there were critical signs showing that the main road curved to the right, a hazardous condition is created for the driver and "run-off-the-road accidents are likely to occur. The function of Positive Guidance at such a location is to change driver expectation by obstructing the visual line of sight or by the use of strategically placed warning signs or delineators showing alignment." A solid statistical analysis showed that chevron signs, amongst other delineators had a significant effect in reducing single vehicle run off road crashes, especially at night (Srinivasan et al., 2009).

A traffic engineering colleague of mine commented upon reading this chapter that the title "Work Zone Crash" is inappropriate, because although it contributed, the work zone was not the major contributing condition – the bad implementation of the realignment of the highway was the issue.

Similar crashes may have occurred on this site since the one described here (Bahar, 2025). I think she has a point!

References

Blaauw, G. J., & Padmos, P. (1982). *Nighttime visibility of various types of road markings: A study on durability, including conditions of rain, fog, and dew* (SAE Technical Paper No. 820412). SAE International.

Bahar, G. (2025) Personal communication.

Grandjean, E. (1988). *Fitting the task to the man: A textbook of occupational ergonomics.* Taylor & Francis Ltd.

McLean, J. R., & Hoffman, E. R. (1973). The effects of restricted preview on driver steering control and performance. *Human Factors, 15*(4), 421–430.

Olson, P. L., Cleveland, D. E., Fancher, P. S., & Schneider, L. W. (1984). *Parameters affecting stopping sight distance* (Rep. No. UMTRI-84-15). University of Michigan Transportation Research Institute, NCHRP Project 1508.

Post, T. J., Alexander, G. J., & Lunenfeld, H. (1981). *A users' guide to positive guidance* (2nd ed.). Federal Highway Administration, U.S. Department of Transportation. Report No. FHWA-TO-81-1.

Smiley, A., & Dewar, R. E. (2016). Road users. In L. R. Pline (Ed.), *Traffic engineering handbook* (7th ed., pp. 51–107). Wiley Publishing.

Smiley, A., Smahel, T., Courage, C., Fitch, G., & Currie, M. (2011). Required letter height for street name signs: An on-road study. *ITE Journal, 81*(7).

Srinivasan R., Baek J., Carter D., Persaud B., Lyon C., Eccles K., Gross F., Lefler, N. (2009). *Safety evaluation of improved curve delineation.* Federal Highway Administration Publication No. FHWA-HRT-09-045. https://www.fhwa.dot.gov/publications/research/safety/09045/09045.pdf

11

Car Versus Bicycle

Introduction

A driver intending to turn left at an intersection struck a bicyclist in front of him. The bicyclist was waiting, also intending to turn left at that same intersection. In the driver's case, as he approached the crosswalk to turn left, his attention was drawn to a vehicle pulled over at the curb to his right. There was also a pedestrian that was partially blocked from his view by the pulled-over vehicle. The bicyclist was equipped with various reflective devices. Despite this, the left-turning driver struck the bicyclist. The Science section discusses visibility and conspicuity, rain, glare and streetlighting, as well as driver attention and information processing.

The Collision

The collision took place at approximately 6:00 p.m. in November. Henry Hunt, aged 46, was driving a BMW Sedan northbound on Crescent Road. David Summers was riding a dark green road bicycle and was also heading northbound on Crescent Road. Mr. Summers had stopped his bicycle in the middle of the intersection with Gothic Street and was waiting to make a left turn. He was then struck by the BMW operated by Mr Hunt. The collision occurred in darkness with overhead streetlights partially illuminating the area. The road surface was wet from rainfall earlier in the day. In the area of the collision, Crescent Road had one northbound traffic lane and one southbound traffic lane. There was a bicycle lane marked in both the northbound and southbound lanes. The posted speed limit was 50 km/h (31 mph). Based on measurements taken at the scene, the police calculated that the Hunt vehicle was travelling at a speed of approximately 40 km/h (approximately 25 mph) at impact.

In Mr Hunt's case, when he approached the crosswalk, his attention was drawn to a vehicle to his right that was pulled over at the curb, and to a

DOI: 10.1201/9781003564805-11

passenger who was alighting from the stopped vehicle. The now pedestrian was partially blocked from his view by the pulled-over vehicle. Just prior to the collision with the bicyclist, Mr Hunt told police that he was concerned about whether the pulled-over vehicle was about to pull into his lane. He noted that he was trying to determine whether the pedestrian was on the sidewalk or on the crosswalk.

The Investigation

I visited the site on two separate evenings. Once was on a Friday and once was on a Saturday in mid-April. For both visits, it was fully dark. The collision had occurred at about 6:00 p.m., 1 hour and 25 minutes after sunset (4:35 p.m.) in mid-November, at a time when it was fully dark. During my visit in April, the sunset was later in the evening (6:07 p.m.) than it would have been in November. Therefore, the equivalent time of the collision in terms of the same light level was later (6:07 p.m. + 1 hr 25 min = 7:32 p.m.). Since it was fully dark at the time of the collision, the visibility measurements could have been taken any time after 7:07 p.m. on the day of my April site visit.

Since it was fully dark when the collision occurred, Mr Hunt would have had to have seen Mr Summers by means of his vehicle's low beam headlights, and by the streetlighting that was present.

I drove northbound towards Gothic Street as Mr Hunt had done just prior to the collision. The driving task is such that when the traffic is moving, a driver's attention would be expected to be directly ahead along the roadway path, and towards any oncoming vehicles.

On Friday evening, the weather was overcast, and the roads were dry. Through the area of the collision, Crescent Road is residential, and a northbound driver has the right of way through a succession of minor intersections, one of which is Gothic Street, the collision site. At the Gothic Street crossing there is a continuous flasher over the crosswalk, as at the time of the collision. A check with the City's traffic engineer revealed that, since the date of the collision, several features that improve information for drivers and improve visibility had been added to the Gothic crosswalk. These included: flashing lights that are now activated by a pedestrian button, two lit median posts on the crosswalk, and two bright streetlights above the crosswalk in addition to the "standard" streetlamp on the west side that appears to be similar to streetlamps at the crossings north of Gothic Street.

As a result of all these changes, I selected an intersection three blocks north of Gothic Street at Simpson Street. The Simpson Street intersection was more like the Gothic intersection, as it appeared at the time of the collision, than to the Gothic intersection in its present modified state.

As was the case for the Gothic intersection, there was a streetlight on the left, for a northbound vehicle, as would have been the case when Mr Summers was approached by Mr Hunt. The streetlight was like other streetlights north and south of the Gothic Street crossing, casting much less light than the two bright lights now at the crosswalk where the collision occurred. Using a Cal-LIGHT 400 calibrated precision light metre, I took light readings on the roadway where the bicyclist had waited to turn left. I aligned the metre's sensor in a horizontal position and then in a vertical position. The light level measured 12 lux in the horizontal position and 19 lux in the vertical position. The adequacy of this is discussed in The Science section below.

I positioned an assistant at the centre line, directly in line with the streetlight, as Mr Summers would have been while he was standing waiting to turn left. I drove a Toyota Yaris, with headlight height like Mr Hunt's BMW, and made observations from the northbound lane at a distance equivalent to 3–4 seconds at 40 km/h (25 mph) from the crosswalk. The reason for choosing 3–4 seconds is as follows. At the equivalent to 3 seconds distance Mr Hunt would have had just enough time to stop from 40 km/h (25 mph). This assumes that his perception reaction time was at the low end of normal (about 2 seconds) and he could brake to a stop within a distance equivalent to 1 second's travel which is 11 m (36 ft). This would require braking of 0.5 g such as is found for drivers braking hard at traffic lights.

According to the police report, at the time of the accident, Mr Summers was wearing the following attire: blue jeans, running shoes, a dark blue bicycle helmet, black gloves, reflective bands which he wore around his ankles, a yellow jacket, and a fanny pack around his waist under his jacket. He also had a green canvas bag which was worn over his left shoulder. The bag hung down on his left side or was worn around him such that the bag was sitting on his back. Police reported that neither Mr Summers nor his bicycle displayed a red rear light or red rear reflector at the time of the collision.

My assistant was dressed in the same jacket that Mr Summers was wearing at the time of the collision. He also had the same dark green bag over his right shoulder. I was instructed by the lawyer for whom I was working, to view my assistant bicyclist with and without a red flashing light. The retroreflective stripes on the jacket could be seen and were not obscured by the green bag. The stripe was conspicuous (i.e. attention-getting) when viewed head on, but **not when viewed peripherally (from the side)**. The flashing red light was conspicuous even in peripheral vision. The yellow jacket itself looked greyish, like a lighter patch of pavement, and the thin white retroreflective stripes were evident. At one point the green bag slid to my assistant's side and the flashing red light attached to the bag was blocked from my view. Mr Summers reported that this had also happened to him. Like Mr Hunt, I saw my assistant against a background of streetlights and oncoming headlights.

In the evening, Crescent Road has a steady traffic flow. Observations had to be made quickly to avoid endangering my assistant. I repeated my observations, but with my assistant on a road bicycle similar to that used by Mr Summers. I made my observations at the Simpson Road intersection which had similar streetlighting and less traffic than the Gothic Street intersection where the crash actually occurred. Observations were easier to make at the Simpson Road location; visibility and conspicuity (attention-getting quality) of my assistant on the bicycle were the same as at the original location.

To help determine if Mr Hunt was driving cautiously, I wanted to know how Mr Hunt's speed compared to other drivers. Using a radar gun, I measured the speed of 15 northbound light vehicles on Crescent Road at Gothic Street at three different points in time: when it was light, but raining steadily (5:00 p.m.), just as it was becoming dark (6:15 p.m.), and again at 7:00 p.m. (about one hour after sunset when it was fully dark). In the latter two cases the roads were wet, but it was not raining.

What do the speed results tell us? Measurements at the site indicate little variation across the day in terms of vehicle speed or ranges of speed. Also, there was little difference when it was raining, compared to when the road surface was wet, but not raining. Based on skid marks and rest positions, the police calculated that the Hunt vehicle was travelling at a speed of approximately 40 km/h (approximately 25 mph) at impact. This is 10 km/h (6.3 mph) below mean traffic speed of 50 km/h (31 km/h). Thus, Mr Hunt appeared to be driving cautiously.

When taking speed measurements, it is important to ensure the radar gun cannot be seen by drivers whose speed is being measured. If possible, readings should be taken after the vehicle has passed by, when the driver can no longer see. This avoids middle finger gestures, but more importantly avoids influencing the drivers to slow down more than they normally would. I once had a case in New Jersey where I had great difficulty convincing a group of uniformed police that were attending the site visit to get out of sight, patrol cars included, when I was attempting to take speed measurements.

The Science

The scientific issues which were important in this case concern driver search behaviour, visibility and conspicuity, limitations of peripheral vision, rainfall and glare effects on visibility, driver information processing, attention, and expectancy while driving. Would most reasonably attentive drivers have noticed the bicyclist, equipped as he was with a yellow jacket with retroreflective strips? Or was Mr Hunt distracted or inattentive?

Visibility

Was the intersection lighting sufficient for Mr Hunt to see the bicyclist? A study of seeing distance to pedestrians at a crosswalk measured visibility for two commonly used lamp types, in the presence of glare. The authors concluded that "a lighting design level of 20 vertical lux is likely adequate for proper pedestrian visibility" (Hankey & Gibbons, 2007). During my site visit, I took readings of the light from the streetlight at the location where Mr Summers had waited to turn left. With the sensor vertical, I recorded 19 lux. Based on the Hankey & Gibbons study, this would have been quite sufficient for drivers such as Mr Hunt to see Mr Summers in time to avoid a crash. In addition to the streetlighting, Mr Hunt would have been aided by headlighting.

Visibility and Conspicuity

Even though it seems to be a low speed, a driver travelling at 40 km/h (25 mph) is moving through space very rapidly – about 11 m (36 ft) every second. The targets that attract attention are objects that are near the drivers' line of sight (within 10–15°), objects that differ greatly from their backgrounds in terms of brightness, colour, texture, etc., objects of large size, and objects that are moving or flashing (Olson et al., 2010).

A study of conspicuity by Wertheim (2010) distinguishes between visibility and conspicuity as follows: "Visibility normally refers to the intensity or salience of an object's features or properties, such as its colour, luminance, form, size, etc. In contrast, the conspicuity of an object refers to how well it visually stands out from its environment (Wertheim, 2010). Hence, conspicuity describes a relationship, the degree to which the object is visually embedded in its surrounding scene. A bright red car can be clearly visible. However, that same car can be either very conspicuous (e.g. when parked in a large green meadow) or very inconspicuous (when parked in a car park with many other red cars)." The same is true for lights and reflections on a roadway at night. A white light or reflection that is conspicuous on its own, will not be when surrounded by other similar lights or reflections.

What reflective elements were there that would have allowed Mr Hunt to detect the bicyclist? How well did they stand out against the background?

Retroreflective Materials

The bicyclist struck in this case was outfitted with retroreflective gear (stripes on a yellow jacket) which would have been lit up by Mr Hunt's headlights. Reflective materials reflect light in all directions whereas retroreflective materials concentrate incident light, that is, light falling on a surface so that most of it is reflected back to the light source. On the roads, this is typically headlights (Tyrrell et al., 2016).

In a discussion of streetlighting, Olson and Farber note that the streetlamps produce visual clutter which "has the effect of reducing target conspicuity, because the target must compete with other elements in the visual field, some of which may be substantially more conspicuous than the target itself" (Olson & Farber, 2003). Streetlamps, headlights, taillights and flashing rear red lights all have light producing elements, and taillights are brighter and therefore more conspicuous than a retroreflective jacket.

At the site visit I noted that the yellow jacket looked greyish, like a lighter patch of pavement. The thin white retroreflective stripes on the jacket were conspicuous (i.e. attention-getting) when viewed head on, **but not when viewed peripherally (from the side) as they would have been had Mr Hunt's line of sight been as shown in Figure 11.1**. The flashing red light was conspicuous, even in peripheral vision. However, it appeared from his own evidence that the green bag carried by Mr Summers had slid to the bicyclist's side, blocking the view of the flashing red light attached to the bag. This also happened to my assistant during the re-enactment.

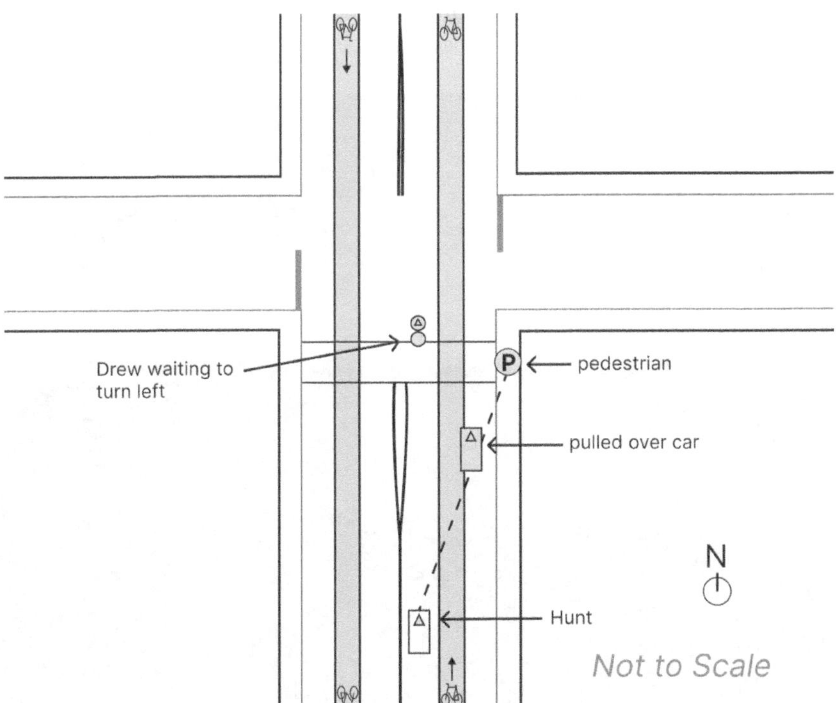

FIGURE 11.1
Shows Mr Hunt's (at-fault driver) line of sight to the pedestrian who was the passenger who alighted from the pulled-over car and may have been about to use the crosswalk.

When I viewed my assistant on the road, his background was streetlights and oncoming headlights on a wet roadway. Figure 11.2 shows visual clutter in a residential neighbourhood at night.

Although we have a wide field of view, the area of accurate vision for the eye is a fairly narrow one called foveal vision. It is 2–4° off the line of sight. How close was Mr Hunt's line of sight to the bicyclist and why did that matter? The last point at which Mr Hunt could detect the bicyclist and stop or swerve would be on the order of 3 seconds' travel distance from the cyclist. At this point, a little high school geometry would tell you that the bicyclist would be viewed 11° off the line of sight, well out of foveal accurate vision. As he approached the intersection, Mr Hunt was looking at the pulled-over vehicle and trying to determine what the pedestrian/passenger was doing with respect to the crosswalk. As Mr Hunt got closer, he further off Hunt's line of sight the cyclist would appear, and the less conspicuous or attention-getting he would be.

Rain and Visibility

According to Mr. Hunt, there was a light rain and his wipers were on "intermittent" setting. In their book Forensic Aspects of Driver Perception and Response (p.109), Olson and Farber discuss the impacts of rain, fog, and

FIGURE 11.2
Visual clutter in a residential neighbourhood.

smoke on visibility. They note that all these elements have in common the property that they absorb and scatter light to some degree and that this absorption and scattering has two effects. "First, less light from the vehicle's headlamps reaches a target object, and less of the light reflected by the object is returned to the driver's eyes. This reduces the object's contrast, making it more difficult to see. Second, some of the scattered illumination is reflected back into the driver's eyes, causing the atmosphere to appear to light up."

Glare and Visibility

The fact that Mr Summers was waiting to turn left tells us that there must have been opposing traffic. Mr Hunt's visibility of Mr Summers would have been reduced by the opposing glare produced by oncoming headlights from southbound motorists. Glare reduces seeing distance because it effectively reduces the contrast of the object being viewed against its background. According to Olson and Farber, "Glare is normally not an important issue with fixed lighting sources, since they are typically far removed from the areas in which target objects occur. One exception is when the pavement is wet (as it was in this case). Not only is pavement luminance greatly reduced, but reflections of the light off the pavement constitute significant glare sources close to the target" (Olson & Farber, 2003). In other words, although the streetlighting would help illuminate Mr Summers, this benefit is offset on wet roads by the glare created when streetlamps are reflected by the wet pavement.

Driver Expectancy

An important factor that must be considered in determining seeing distance is driver expectation. Whether or not a driver is expecting to see something in the road has a great impact on seeing distance. Knowing the nature of the target makes it possible to use subtle cues to detect it further away than might be possible under normal conditions.

Although there was an overhead flashing light drawing attention to the crossing, it was continuous, rather than being activated by pedestrians desiring to cross. As this was a very familiar route to Mr Hunt, the continuous nature of the light did not provide any new information about the presence of a crosswalk. The additional activated light, now at the crosswalk, was not present at the time of the collision. Therefore, the pedestrian who was in fact trying to cross could not provide drivers with a definitive signal of her intention, as is possible today, by activating the additional flashing light.

There was a bicycle path along the right side of the northbound lane, which should lead drivers to expect bicyclists. However, Mr Hunt may not have expected a bicyclist at the intersection, in the left side of the through lane, although this was the correct move on Mr Summers' part.

Driver's Visual Search Patterns While Driving

The visual field of the human eye is large, with only a small area of the visual field, called the fovea, allowing accurate vision. The quality of vision falls off rapidly away from the fovea. For this reason, in order to identify a target, we look directly at it. We search the visual scene in a series of fixations, fixating on successive objects of interest. Studies of driver eye movements show that, when stopping and entering an intersection while driving, fixations range from 1.1 to 2.6 seconds (for checking right and left for traffic in the presence of traffic in one of or both the left and right directions). It is likely that the shorter fixations (1.1 seconds) were made when drivers were moving more quickly. It only takes a few seconds to traverse an intersection. This means that the number of fixations that are made, and the number of objects that can be identified as a driver stops or slows to check for traffic on the left and right, is very limited. The more complex the visual scene, the more visual searches there are, and, to a lesser extent, the longer the individual searches are.

In Mr Hunt's case, as he approached the crosswalk his attention was drawn to a vehicle pulled over at the curb to his right, and to a pedestrian who was partially blocked from his view by the pulled-over vehicle. Just prior to the collision with Mr Summers, Mr Hunt was concerned about two issues: 1) whether the pulled-over vehicle was about to pull into his lane, and 2) whether the pedestrian/passenger was on the sidewalk or on the crosswalk, and about to cross in front of him. He certainly could have been looking away from the forward view for 1–2 seconds at a time.

Attention, Information Processing and Driving

Human attention and abilities in information processing are limited. These limitations can create difficulties because driving requires the division of attention between control tasks (e.g. staying in the lane), guidance tasks (e.g. merging with other vehicles), and navigational tasks (e.g. looking for street name signs). While attention can be switched rapidly from one information source to another, humans only attend well to one source at a time. Information sources include such things as the activity of an oncoming car, state of a traffic light, street name sign, and so on.

Drivers can only extract a small proportion of the available information from the road scene. It has been estimated that out of over 1 billion bits of information per second directed at the sensory system, on average, we consciously recognize only 16 bits of information. (The answer to a yes/no question provides 1 bit of information.) Thus, in some situations involving divided attention, key information will be missed. A naturalistic on-road driving study involving the instrumentation of 100 vehicles over a 1-year period found that 19% of crashes occurred when drivers glanced away from the forward point of view, due to another driving task demand (Dingus et al., 2006). Thus driving-related "inattention", due to our limitations in dividing attention and processing information is relatively common.

Initial research suggested that the human information processing system was essentially a single channel system with limited capacity (Grandjean, 1988; Kantowitz & Sorkin, 1983, Olson et al., 2010). In other words, humans could only process one channel of information at a time. However, experience shows us that we can talk to passengers, watch for other vehicles and for signs and signals and still drive millions of miles without a collision. The reason is that humans can switch their attention rapidly between tasks. However, they have a limited number of resources and the system only "works so long as the combined demands stay within the resource capabilities of the operator" (Olson et al., 2010).

Collisions resulting from high demands from more than one information source were recorded in a study carried out by Virginia Technical Transportation Institute for the U.S. National Highway Safety Administration. This study involved the instrumentation of 100 vehicles so that naturalistic driving over a 1-year period could be recorded from both principal and secondary drivers (241 in total). Approximately 20 million miles of driving were recorded. Video images were taken of the road scene, before and after, and of the driver's face. Speed, acceleration, brake pressure, and lane position, among other variables, were continuously measured. During the course of the study, 69 crashes with full data, 761 near-crashes and 8,295 incidents were recorded.

Dingus et al., (2006) defined inattention as glances away from the forward view occurring within 3 seconds of the conflict or crash. Several sources of inattention were identified: secondary tasks (e.g. using a cell phone, looking at an attractive pedestrian on the sidewalk, etc.), drowsiness, and non-specific reasons (glances away from the forward view without obvious motivation). In addition, researchers described a new category of "inattention", namely driving-related inattention. This latter category includes making glances away from the forward roadway due to checking the speedometer, checking blind spots, looking for a parking place, or changing lanes, at an inopportune time. About 19% of crashes occurred when drivers glanced away from the forward point of view due to another driving task demand.

Discussion

Mr Hunt was faced simultaneously with several driving task demands. Directly ahead of him was Mr Summers who was stopped at the intersection in the through lane. To his right was a vehicle which Mr Hunt thought might pull back into traffic in front of him. Furthermore, there was a pedestrian behind the vehicle who might have been on the crosswalk. Mr Hunt's view of her was blocked by the vehicle from which she emerged. In response to this uncertainty, Mr Hunt, based on police calculations, appeared to be driving cautiously, at an estimated 38–42 km/h (24–26 mph), comparable to the low end of speeds that I recorded on this stretch of road both in rain and in dark conditions.

Given the presence of headlights and adequate streetlights, combined with the retroreflective materials on Mr Summers' jacket, there is no question that Mr Hunt would have been able to see Mr Summers' striped jacket. However, this would assume that he looked in that direction. He could have seen this well in advance of the distance he would need to respond to Mr Summers' presence by braking to a stop. The question is whether Mr Hunt would have noticed the bicyclist given the circumstances. Although we have a wide field of view, the area of accurate vision for the eye is a narrow one called foveal vision. As previously explained, foveal vision covers 2–4°, about the size of a quarter held at arm's length. Beyond 4° is peripheral vision. The further a target is from the line of sight, the less likely it is to be detected.

If Mr Hunt was looking at the vehicle pulled over at the crosswalk, trying to determine whether the pedestrian was on the sidewalk or on the crosswalk, and whether or not the vehicle was about to pull into his lane, he would have been doing a difficult visual task using his high-resolution foveal vision.

If Mr Hunt was looking towards the pulled over vehicle and pedestrian as he approached the intersection, they would be seen along his line of sight. Therefore, he would have had to detect Mr Summers in peripheral vision. How far into peripheral vision would that be? In other words, how difficult would it be to detect Mr Summers? Drivers must look ahead in order to keep within their lane. Two seconds of preview (i.e. travel distance) is found to be the bare minimum for this (McLean & Hoffmann, 1973). Drivers must also look far enough ahead to be able to respond in time to changes in the road path and/or hazards such as a jogger suddenly crossing the intersection in front of them or a large pothole. As explained in the Investigation section, 3 seconds travel time at least 40 km/h (11m/s) allows for travel during PRT (2 seconds of travel) and travel distance (1 second equivalent) sufficient to complete an avoidance manoeuvre, assuming hard braking.

At a travel distance equivalent to 3 seconds away from Mr Summers, if Mr Hunt was looking at the pulled-over vehicle and the pedestrian, Mr Summers would have been seen in his (Mr Hunt's) peripheral vision, 11° off his line of sight. As Mr Hunt came closer, this angle to Mr Summers would increase making it more difficult for him (Mr Hunt) to detect the cyclist if he continued to look towards the pulled-over vehicle and its passenger. A study of driver ability to detect disc targets on sign posts near the roadway during the day showed that the majority were noticed when they were less than 10° off the line of sight (Cole & Hughes, 1984).

At the time of the collision, the level of lighting from the streetlight and low beam headlights was adequate for Mr Hunt to detect the bicyclist, had he been looking straight down the road ahead. However, there was light rain, and the windshield wipers were operating intermittently. There was direct glare from the headlights of opposing vehicles, as well as reflected glare from the wet road due to both headlights and streetlamps further up the road. The opposing headlights would be considerably brighter than Mr Summers' clothing would have been, even with some retroreflective striping on his coat and ankles. In these circumstances, Mr Summers would be visible for a driver who looked straight ahead at his path into the intersection. However, he would not have been conspicuous when seen in peripheral vision.

Tragically, Mr Hunt was approaching Mr Summers at the same time that there was a vehicle stopped at the crosswalk letting off a passenger. Given the complexity of the situation, Mr Hunt's glances could easily have lasted for 2 seconds or more as he closed in on the bicyclist. Due to his attention to the pulled-over vehicle and the pedestrian to his right, he failed to detect Mr Summers in peripheral vision to his left. This was despite the fact that Mr Summers was wearing a jacket with retroreflective stripes.

If Mr Summers had also been equipped with a red flasher which was visible (not having slipped to the side), then Mr Hunt should have detected that in peripheral vision, since a flashing light would have substantially increased his conspicuity in peripheral vision.

In addition to the stripes on his jacket, Mr Summers was also wearing retroreflective bands around his ankles. These are very effective at catching driver's attention when the ankles are moving, however in this case Mr Summers was waiting in the intersection, not moving.

Conclusion

In my opinion Mr Hunt bore the main responsibility for this collision. In his defence however, he was faced with several driving task demands simultaneously, resulting in his attending to a pulled-over vehicle and a pedestrian potentially on the crosswalk to his right in the very seconds in which he needed to detect Mr Summers directly ahead. Despite his yellow jacket with retroreflective striping, Mr Summers would not have been conspicuous in a background of bright streetlamps and headlights, when seen in peripheral vision. In my expert opinion, had Mr Summers' bicycle been equipped with a red flashing light visible from the rear, this would have substantially increased the likelihood of him being detected in peripheral vision in the last few seconds before impact.

Countermeasures

A continuous flashing light over the crosswalk tells drivers where the crosswalk is, but nothing about pedestrian presence. An overhead flashing light activated by the pedestrian, now at the crosswalk, was not present at the time of the collision. Had it been present, and, very important, activated by the pedestrian, it might well have prevented this collision by helping the driver to quickly determine that there was a pedestrian in the crosswalk, allowing the driver to shift his attention back to the view ahead where the bicyclist was waiting to turn. However, the driver still has to look because they cannot rely on the pedestrian activating the light.

References

Cole, B. L. & Hughes, P. K. (1984). A field trial of attention and search conspicuity. *Human Factors, 26*(3), 299–313.

Dingus, T. A., Klauer, S. G., Neale, V. L., Petersen, A., Lee, S. E., Sudweeks, J., Perez, M. A., Hankey, J., Ramsey, D., Gupta, S., Bucher, C., Doerzaph, Z. R., Jermeland, J., & Knipling, R. (2006). *The 100-car naturalistic driving study. Phase II – Results of the 100-car field experiment* (Report No. DOT HS 810 593), National Highway Traffic Safety Administration (NHTSA).

Grandjean, E. (1988). *Fitting the task to the man: A textbook of occupational ergonomics.* Taylor & Francis Ltd.

Gibbons, R. B., & Hankey, J. (2007). Wet night visibility of pavement markings: Dynamic experiment. *Transportation Research Record, 2015*(1), 73–80.

Kantowitz, B., & Sorkin, R. D. (1983). *Human factors: Understanding people-system relationships.* Wiley.

McLean, J. R., & Hoffmann, E. R. (1973). The effects of restricted preview on driver steering control and performance. *Human Factors, 15*(4), 421–430.

Olson, P. L. & Farber, E. (2003). *Forensic aspects of driver perception and response* (2nd ed.). Lawyers & Judges Publishing Company.

Olson, P. L., Dewar, R., & Farber, E. (2010). *Forensic aspects of driver perception and response* (3rd ed.). Lawyers & Judges Publishing Company.

Tyrrell, R. A., Wood, J. M., Owens, D. A., Whetsel Borzendowski, S., & Stafford Sewall, A. (2016). The conspicuity of pedestrians at night: A review. *Clinical and Experimental Optometry, 99*(5), 425–434.

Wertheim, A. H. (2010). Visual conspicuity: A new simple standard, its reliability, validity and applicability. *Ergonomics, 53*(3), 421–442.

12

Acting as a Human Factors Expert Witness

Introduction

In this chapter, I provide insight into work as a human factors expert witness. I begin with how one finds work in this area, then describe the various stages involved in a legal case, including the initial contact, the site visit, writing of a report, preparation for court, qualification as an expert in court, the examination-in-chief, the cross-examination, the re-direct, and the aftermath. Throughout I use illustrations from my own experiences and expertise an expert witness in human factors.

Finding Work as an Expert Witness

How does one find legal work? Typically, it finds you. Usually a lawyer for one side of the case phones you. He or she will have identified you as a potential expert from talking to colleagues, or identified you from your publications. The lawyer may have had a literature search done to determine who might have expertise in the particular area of interest – for example, the effects of talking on a cell phone while driving, or marihuana effects on detecting a hazard while driving. Some experts get themselves listed on websites identifying expert witnesses in various areas. The company that does this collects a substantial cut of the proceeds. In short, there are a number of ways that legal work finds the relevant experts.

Qualifications for Being an Expert Witness

When the court calls on experts, the lawyers, judges, and juries must be able to rely on the expert's words, because they are not in a position to dispute

DOI: 10.1201/9781003564805-12

what an expert says on technical grounds. Whether or not an expert is adequately qualified is a serious matter. It is dealt with by means of a formal qualification process in court which takes place before one is allowed to give evidence (more on this later). I must add that I have twice seen engineers with no human factors training give human factors evidence in a legal case. However, this is risky because an informed lawyer could easily challenge their "expertise." I once waited two hours for a judge to decide whether I had the qualifications to give evidence about PRT. The decision to accept me as an expert was made based on a series of questions to me concerning the content of my various courses in university, many years previous, and how that content related to perception reaction time. I was able to remember what courses and topics I must have studied, that I covered in my legal report. My ability to connect the issues I was discussing with the content of those courses was the basis of a decision by the judge to allow me to go ahead and testify.

Initial Contact

Suppose a lawyer has contacted you. The first thing he or she will want to know is whether you have any conflict of interest. In other words, have you already been already hired on the other side of the case. This is determined by providing the date of the crash and the names of the parties involved. Once it is established that you have no conflict of interest, the lawyer will then provide you with the details of the collision.

If you are satisfied that the issues involved fall within your area of expertise, you then can feel free to take on the case. The lawyer will first send you a bundle of documents for review. In the early days of my career as an expert witness this bundle of documents came in several cardboard boxes. Nowadays the material mostly comes electronically, making it easier to share with other people working on the file. These documents include statements to the police of the involved parties (e.g. the driver at fault, the pedestrian struck) and witnesses (e.g. a passenger), exams for discovery of the involved parties, a transcript of a preliminary inquiry, investigator reports, reconstruction engineering reports including Electronic Data Records, highway design reports, forensic reports (e.g. blood alcohol levels, condition of headlights: lighted or not at the time of the crash depending on the filament), manufacturer's product description (e.g. of a motorcycle), photographs of the crash and/or the crash scene at some later point in time, any videos that show the collision, including those from the involved vehicles if available, and those from security cameras the collision site if relevant. In addition an expert witness will often access Google Earth and Streetview records.

In the letter accompanying these documents, many lawyers will provide instructions concerning what questions they would like you to answer – For example, for a collision involving a pedestrian, the instruction might be:

> "Assume that Ms G did not have her turn signal on before she slowed quickly and turned left. How, if at all, does this affect your opinion of whether the following driver was too close?"

Many lawyers do not provide letters of instruction but simply ask for a human factors report, trusting you to select the issues of importance.

One must keep careful records of these cases because years go by between contact and actually doing any work on the case. I once was reading through initial documents, thinking to myself that the crash details seemed familiar. I did some digging and found to my horror that I had been hired on the other side the previous year! The identifying information for the first case was missing the date of the crash and used some different plaintiff names. On the advice of the first lawyer who hired me I immediately shipped back all the documents to the second lawyer. Ironically, the second lawyer was able to use my evidence to win his side of the case. The judge complimented me on being independent in my opinions, and thus, in his view, a true expert witness to assist the court.

Site Visit

When dealing with traffic collisions, it is usual to go to the site. Even after reading over all the reports and looking at the photographs, there are still things to be discovered at the site itself. One case of mine involved a pick-up truck without functioning brake lights but with an operating turn signal, which was rear-ended when it turned off a rural highway into a narrow farm laneway. The driver behind presumably saw the turn signal but did not have the benefit of seeing the brake lights come on as the pick-up truck slowed for the turnoff. (As you may now know from Chapter 9: Car Versus Tractor, it is difficult for drivers to assess how quickly they are closing in on the vehicle ahead just by looking at the change of image size of the vehicle ahead.) To avoid the pick-up truck in front of him when it stated braking hard, the surprised following driver swerved into the next lane and had a head-on collision with an oncoming car. Upon visiting the site I discovered that there were a number of laneways within a short distance, making it difficult for the following driver to guess which lane the pick-up truck in front was about to turn into once he saw the turn signal. I also discovered that the laneway was very narrow. I had to slow down considerably from the speed limit in order to enter it.

A visit to the crash site should take place at the same time of day, same day of the week, same season, and under the same weather conditions that the crash occurred. Lighting, traffic density, and road surface conditions are all affected by these variables. It is helpful to bring to the crash site the police report and other documents so that the point of impact can be established. An expert in human factors issues should see the site in as close a condition as possible to that of the person involved in the crash. One should approach the crash scene along the same route as the person involved. As much as possible, one should see things the way the driver or pedestrian did just prior to the crash.

One tricky issue can be determining the time to visit the site that mirrors the level of light that existed at the time of the crash. For example, let us suppose that the crash occurred at 7:37 p.m. on September 15th, 2022, at 10 minutes after sunset, when light levels are changing rapidly. Let us also suppose that you cannot visit the site until September 25th, 2022. Sunset on September 25th, occurs at 7:09 p.m., 22 minutes later than on September 15th. On September 25th, 10 minutes after sunset is 7:19 p.m. This is the equivalent time to 7:37 p.m. at which you must record visibility measures.

An expert must be particularly careful with determining the equivalent time for crashes that occur around sunset or sunrise and around the time of the year that we change our clocks. It is easy to get it wrong and select the wrong equivalent time, as I did on one case. Fortunately, I had to see the site on a second occasion and so I was able to see it at the right time and in the right lighting and was not forced to reveal my error to the lawyer on the case. Mistakes like this are embarrassing. A colleague made such a mistake and saw the site an hour later than he should have. Although he admitted it, he claimed that it did not make any difference to the visibility of the vehicles involved. This was so obviously biased by his concern about admitting his error that his evidence about visibility would have been considered with some doubt or questions, had it been presented in court.

Another example of the need for precision in seeing the site under similiar lighting is the case I had involving a truck driver's ability to see red flashing lights at a rail crossing in full sun. The flashers were covered with dried salt spray. The driver claimed he could not see the flashing lights, providing a reason for his ending up driving into a train. The crossing was 1,000 km from my home. Mindful of the considerable price of the flight, at 6 a.m. I checked with the air traffic controllers for the expected weather conditions at my destination before setting out. "No problem" they assured me. "There is a little fog now, but it will burn off and be long gone before you get here." No such luck! There was so much fog we were not only late, but we could not even land at the ticketed destination. Instead, we had to land at an airport far enough away that, despite quickly renting a vehicle to drive to the crash site, I was not able to meet the noon hour deadline. I had to fly back again two weeks later, doubling the cost to my client. And, my findings about the visibility of the flashers at the train crossing did not support the rail company that hired me.

Photographs of the collision site are important because they will help establish whether the scene has been altered since the crash. It is common to be examining crash scenes several years after the crash has occurred, particularly in civil cases. It has been my experience that traffic signs, road design elements, or lighting are often changed or added in the aftermath of a serious crash. The change can be sufficient enough to alter subsequent driver behaviour.

There may be measurements that will be helpful in developing an opinion on what happened. The equipment we normally take with us on a site visit can include a set of safety glasses with a mini camera embedded between the eyes (to record the driver's view), a radar gun for sampling speeds, and an engineering wheel for measuring distances. When investigating a night-time crash, you would take fluorescent traffic cones and safety vests. Investigating a collision provides a potent reminder that carelessness about safety can lead to tragic consequences.

After accepting a case, one should be careful about discussing it. I often consult my international colleagues in traffic safety. When I first started working as an expert witness, I had a case involving the perception of closing speed: when drivers perceive that they are closing rapidly on a vehicle in front. I contacted the expert I believed to be pre-eminent in the world on this particular issue. I began discussing the case and, to my horror, he said, "Alison I don't think we should discuss this matter further. I believe I have been hired by the opposing side." That information made me even more stressed out than usual about going to court.

Typical human factors issues which are raised in traffic collision cases include the following: drivers hitting pedestrians at night, with one or other or both being impaired; drivers hitting unexpected and unmarked obstacles on the road at night; drivers rear-ending a vehicle in front which was going much more slowly than the driver expected; drivers running a stop sign or a red traffic light and hitting a vehicle crossing the intersection; tired truck drivers who rear-end another vehicle; and motorcyclists being hit by drivers crossing their path who do not see or notice them.

To help develop scientific evidence in such cases, I rely on literature obtained using Google Scholar, as well as on my personal collection of reports, publications, and conference proceedings collected over 50 years.

Preparing a Report

Often your lawyer client will ask for a meeting to determine what you would say *if* you wrote a report. If the lawyer does not think the report is in his or her favour, you may be asked simply to invoice for your time and that will be the last you hear of it. Once the lawyer asks you for a written report, they are obligated to provide that to the other side, and they naturally do not want to do so if it is not in their favour. Occasionally, on very borderline opinions, I have written the opinion and read it out over the phone to the lawyer. In this case, they are under no obligation to provide it to the other side should they not like what it said.

If a written report is required, it is usual to begin with a list of the documents one was given to review, and the questions from the letter of instruction or mandate, followed by a brief description of the crash, using information from the police report. In my reports, I then list the human factors issues and discuss each one in turn. One does not need to provide references in the same manner as in a publication, although many experts do.

At the end of the report, I summarize the preceding discussion and attempt to give a very clear opinion. In a civil case involving a traffic crash, the question is one of ability. One should be able to give an opinion as to whether an average driver ought to have been able to avoid this crash. In a criminal case the issue is usually one of whether a driver acted in a knowingly risky manner (dangerous vs careless driving).

In my opinion, there are cases where criminal charges are brought against drivers inappropriately. The error the driver has made can be a common one. For example, as a human factors expert I am aware that it is impossible for drivers to be continually vigilant. It is an integral part of normal behaviour to be inattentive occasionally. Usually, such inattention does not occur at a critical moment. All drivers have experienced this.

Occasionally, this inattention does occur at a critical moment, for example, just prior to a stop sign. Many, if not most people have had the terrifying experience of unintentionally running a stop sign. This is bad enough, but if it occurs when you are on a collision path with another vehicle, a fatality can result. My experience is that jury members pay rapt attention when I talk about this – they recognize themselves.

In such cases, I have argued that even prudent drivers have such moments of inattention and that while a driver who does this may be termed to have driven in a careless manner, this by no means implies an intentionally risky manner. A careless-driving charge can results in demerit points and other sanctions. A charge of dangerous driving means a criminal record, and a possible jail sentence.

When one writes a report, one must be prepared to defend every statement, as long as new evidence has not changed your mind. Different jurisdictions have different rules. In my jurisdiction, an expert's report must be submitted with a detailed resume three months prior to the start of the trial. Trials often go on for several weeks. One needs to keep in mind that lawyers and experts on the other side will therefore have a considerable amount of time to go through all elements of your report.

In particular, care should be taken not to comment on matters outside one's expertise even though one may have a good understanding of them. For example, I have avoided commenting on braking distances for trucks or motorcycles, with which I am familiar with but not an expert in. When necessary I will ask the lawyer to retain another expert to provide such information, which I may need to support my opinion.

You may be given a report by an expert on the opposing side of the case. It may even be someone you know well. It is important to extend professional courtesy to the other expert. I well remember an expert's comment on my report. He began by saying that he had read mine and while he fully agreed with me on the first three points I had raised, he must respectfully disagree on the fourth. He went on to state his objection. By being courteous in his treatment of me, he gave credibility to our profession and to himself.

One must keep in mind that experts appearing in court are there to provide scientific evidence and must be as unbiased as possible in presenting it. There should be some overlap, although not necessarily complete agreement between experts on opposing sides of a case. Otherwise one or the other, or both experts, become less credible. In my jurisdiction, experts are required to complete an Acknowledgement of Experts form which says in part:

"I acknowledge that it is my duty to provide evidence … as follows:

1. to provide opinion evidence that is fair, objective and non-partisan;
2. to provide opinion evidence that is related only to matters that are within my area of expertise; and
3. to provide such additional assistance as the court may reasonably require to determine a matter in issue.

I acknowledge that the duty referred to above prevails over any obligation which I may owe to any party by whom or on whose behalf I am engaged."

This is a good reminder.

One must consider very seriously one's position on various issues before putting that opinion in writing or testifying before a court. Let us suppose that in one case, you are hired on the side of a child who runs out between parked cars and you say the driver should have been able to stop in time. You may then be hired to look at a similar case, but this time on the side of a driver. You cannot ethically now say that the driver in the identical situation could not have stopped in time. Nor should you if you want to maintain your credibility as an expert. The lawyer on the opposing side has access to legal websites which can be used to trace what you have said in previous similar cases. He or she also has your resume and can obtain copies of any of your publications that might be relevant. If they contradict your current position, having your own words thrown back in your face while you are on the stand would undermine your credibility.

Sometimes an expert is asked for a verbal opinion on a case and no formal report is written. This happens on the defence side of a criminal case. The Crown must be notified that the defence intends to call an expert witness and the defence must provide the expert's resume to the Crown. A verbal opinion could also be required if the lawyer is unsure about what your evidence might be or is pessimistic about the possibility that your evidence will be helpful to his or her case. If, for example, you are asked to comment on a case involving a driver hitting a child who runs into traffic, you may have to say, given what is known about driver PRT, that the driver could not have avoided the crash. It would be of no value to the lawyer acting for the child to have this opinion in writing. Your involvement in the case will then end at this point.

On one occasion I was on the opposite side of a case from a close colleague. I was tempted to write a joking email, saying "pistols for two, coffee for one," followed by the trial date. However, I found I was on the "bad" side of the case. I knew my opinion would not help my lawyer and so no report would be written or entered into evidence. In this case letting my colleague know I had accepted the case, but there was no legal report from me, would have told his lawyer that he or she had the upper hand. Therefore, I sent no such email.

Sometimes the lawyer will ask for a copy of your draft report. Upon reading it, he or she may request changes. Mostly, depending on how good a writer you are, these changes are not substantive, but sometimes they are (e.g. removing a reference to what you think is a critical contributing factor to a crash, but is not in your client's favour). For example, you can hardly ignore the fact that your client was at twice the legal limit for alcohol. Ignoring this can make you look biased. You must be very careful that you only make changes with which you are fully in agreement. After all, you must defend the report on the stand. You do not want to find yourself explaining why you made the changes you did between one version and the next, for example, whether you initiated the change or the lawyer did.

I once had the occasion to work for a very well-known civil rights organization and one whose mandate I supported fully. The lawyer for said organization asked for a copy of my draft report and inserted a number of changes. I did not agree with a lot of them. Before I could contact her and object she sent another version with complete paragraphs inserted into my report. When I objected she said that she did this for lots of experts and implied that she was a better writer, which was probably true. However they were the ones giving the opinion. I objected to the paragraph she wanted to put into my report. As a result I was fired! This was unfortunate because they had an expert in their corner (me) who was very much inclined to their point of view on the matter in question.

Preparing for Court

Once a trial date has been selected, the lawyer with whom you are working will contact you to let you know the date and to check on your availability. There is usually some flexibility about when you can be called. For example, if you have an important meeting on a particular day, the lawyer can usually work around this. If you will be unavailable for a long period, the lawyer may even reschedule the trial if you are an important witness.

My experience on the matter of appointments to appear in court is that they almost never happen as planned. Trial dates are often rescheduled. Other witnesses take longer than was originally foreseen. Your testimony may take longer than expected. I have even had the experience of flying to another city, waiting to testify throughout one day, returning the following day and, just prior to my planned appearance, being told the lawyers decided against having me testify. It seems I had raised a point not in favour of their client in my report. Since no one until that moment had raised that point, they decided it would be better to "let sleeping dogs lie."

Due to court scheduling, I have been seriously delayed twice: the first time resulting in missing a course I was supposed to teach on traffic safety to 35 state transportation professionals, and the second time I missed out on a weekend in Spain.

Cases can be settled at any time, often before the costly court process has begun, but sometimes during the trial. It is even possible to have the case settled during a recess in your testimony. The majority of the cases in which I have been involved (about 95%) have not required my testimony in court.

Cases get thrown out of court as well. During a conference I attended some time ago, I was prepared on a moment's notice to fly back to another city to testify in a case. As it turned out, the crown witnesses could not agree as to which lane the crash had taken place and the case was dismissed.

Despite knowing all the above, you must assume you will be testifying when called, and prepare accordingly. Before I walk into court, I will have usually gone through my testimony with my lawyer and I will have all the pertinent facts of the case at my fingertips. I will have the details of up to a dozen studies which support my opinion in my head, and I will have all the documentation that I was sent, and copies of all of the publications I relied on in my report, in my possession. This is a usual requirement when appearing for an Examination for Discovery or a trial.

Qualification

A word of warning – for a profession that is all about usability, it is unfortunate that we ergonomists use different nomenclature depending on whether we deal with human cognitive limitations (human factors specialist) or human physical limitations (ergonomist). "Ergonomics" was the term used when our profession was started in Europe from physiological roots; and "human factors" in North America which started in the military. Both the

Canadian and the U.S. professional associations include both terms in the organization name.

In professions with a long history and a well-established qualification process, qualification in a court is relatively simple. No lawyer is going to waste time arguing whether a physician is a qualified doctor.

Legal qualification as an expert in court for a human factors specialist/ergonomist was more complicated before there was a Canadian College of Certified Ergonomists (founded 1999) which has specific educational and practice requirements, a code of ethics, and continuing education requirements. Belonging to a professional certifying organization simplifies the qualification process.

When writing legal reports and in court, I present myself as an expert witness in human factors, specializing in driver behaviour and traffic safety. To give an example of appropriate qualifications in this area, I give my own, versions of which have been accepted in court on numerous occasions:

> I am a Canadian Certified Professional Ergonomist. I have been president of Human Factors North Inc., a Toronto-based human factors engineering consulting company, since 1984. I hold an MASc (1971) and a PhD (1978) in Systems Design Engineering, specializing in Human Factors. I am an adjunct professor in the Department of Mechanical and Industrial Engineering at the University of Toronto where I taught courses in human factors for 30 years. I have served on a number of U.S. Transportation Research Board committees starting in 1978 and continuing to the present, and am an emeritus member of the Vehicle User Characteristics Committee. I serve on the board of the journal *Accident Analysis & Prevention*. I was the 1997 recipient of the AR Lauer Safety Award given by the U.S. Human Factors and Ergonomics Society for outstanding contributions to the human factors aspects of highway safety. Over the past 35 years I have acted as an expert witness in over 500 motor vehicle crash cases, carried out research studies on driver behaviour using simulators and instrumented vehicles, participated in road reconstruction and design projects with numerous engineering companies and developed and taught human factors and road safety courses to graduate and undergraduate students, traffic engineers, highway designers, police and other transportation professionals across Canada and the U.S.

After being sworn in, the lawyer who is presenting you as their expert witness will go through your resume, asking you about your degrees, your employment, and your publications. Your lawyer will often focus on particular reports depending on the case. For example, if vision with headlights is an issue, research done in this area would be emphasized. Your lawyer will ask you to confirm the truth of each item mentioned. It can feel like an extended session of boasting, but as noted above, it is very important that the court assures itself of your expertise before allowing you to give opinions that may sway the case one way or another. For us, who practice a profession not widely understood, it is worthwhile to begin with a short and simple description of what human factors is all about.

After your lawyer has gone through your qualifications, the opposing lawyer has his or her turn. My experience is that the more damaging your testimony is likely to be to their client's case, the more likely he or she is to object to your being accepted as an expert.

A number of times I have had to defend myself against the charge that my evidence had no scientific basis. In a U.S. case, I had an aggressive opposing lawyer, who, upon encountering me accidentally in the hotel hallway said "we are going to eat you for breakfast tomorrow," to which I responded "you will choke." I was to give evidence on circadian rhythm effects on sleepiness. He started out by mocking my evidence, waving around an article on the moon's effects on circadian rhythms, equating them with biorhythms. At this point I realized that the reason he was being so aggressive was that he did not have his own human factors expert and thought he could match me nonetheless. In response to his disdainful tone, I responded that I had with me a publication entitled *Design of Schedules for Air Crew Based on Circadian Rhythm* by NASA. I said "They seem to take circadian rhythms pretty seriously."

A curious aspect of the qualification process is that you will often be asked if you have given testimony in a court before. If you do it all the time, the opposing lawyer may try to present you to the jury as a "hired gun" who is willing to testify in favour of any side that hires you. If you advertise yourself as an expert, this may open you up to questions of bias. The opposing lawyer may ask how much you have been paid per hour, or even how much money you have made on the case. This is calculated to create a negative impression with the jury, few of whom are paid as much per hour for their time. If you have had very little experience in court, then the opposing lawyer may take the tactic that you are less credible because you are inexperienced. I did have a good laugh at a colleague's response to this type of question. When asked if he was not simply a hired gun, his response was, "Oh no. I have not been called to the bar!"

Examination-in-Chief

Once the court (i.e. the judge) has accepted you as an expert in the area in which you propose to testify, the examination-in-chief by your own lawyer will begin. Generally, your lawyer will work from the formal report or from notes you have provided and the two of you will have rehearsed the questions and answers before going into court. In a traffic crash case, the examination often begins with your lawyer asking for your understanding of how the crash happened. He or she will then ask if and when you visited the site of the crash. He or she will then lead you through your report, perhaps asking first what issues you feel are significant and then asking you about each one in turn. Finally, he or she will ask for your opinion. The question may be, "In your opinion, would the average driver have been able to avoid this accident?" Lawyers often ask as to whether a prudent driver would be able to avoid the accident. This is tricky ground. Human factors specialists are qualified to explain why the crash happened, not who is to blame for it. The latter is job of the judge and jury.

The examination-in-chief usually takes place as a series of questions and answers. This is easier for a jury to listen to than a long lecture on your part. When a jury is present, you need to remember to speak directly to them and to avoid jargon that many jury members will not understand. Judges usually make notes and may even ask you to repeat points or slow down so they can get the information down.

During the examination-in-chief, your lawyer may ask you about issues that are damaging to your side. He or she will do this if she knows the opposition is bound to bring the issue up. This way he or she steals the thunder from her/his opponent and by admitting the problem, reduces the damage that revealing the issue will cause.

Your lawyer or the opposing lawyer may ask you for opinions on matters that are outside your area of expertise. Do not be tempted. He or she may do this in order to have you say, "I'm sorry I cannot offer an expert opinion on that. It is outside my area." This will help establish you as a reliable expert.

I once was caught in a quandary about how to answer a lawyer's question on a particular subject. It concerned a driver who had run a stop sign at night after engaging in a one-night stand. The lawyer prosecuting my client said to me, "Doesn't having sex make you feel sleepy." I was not sure whether to say it was outside my area of expertise or frame an answer indicating it was based on my limited experience!

Cross-Examination

The most challenging part of the legal process for experts is the cross-examination. In Canada, it still happens sometimes that the other side does not have a human factors expert witness. When this happens, things are relatively easier for you as compared to when the other side has an equally qualified expert witness who can feed nasty questions to the lawyer cross-examining you. The Appendix to this chapter contains my reactions to the experience of a day on the stand.

Unless the lawyer has another expert present in court feeding difficult questions, the lawyer's main attack will be on your credibility as an expert.

In my testimony before the Commission of Inquiry into the Hinton Rail Disaster, I was asked many questions which were totally irrelevant to my opinion. For example, I was asked if I had visited the signal manufacturers to discuss the reliability of railway signals. I had to answer "no." In another trial, the cross-examining lawyer addressed me as Ms Smiley throughout his examination, despite the judge and my own lawyer having addressed me as Dr Smiley. This same lawyer challenged me when I said that the most frequent traffic convictions were for speeding. Now, lawyers are taught never to ask a question of an expert unless they already know the answer. It can backfire as it did in this case. It so happened that I had just done a study examining 400,000 Ontario driver records and that was the basis for my claim that the most frequent traffic convictions were for speeding. As soon as I said "400,000 Ontario driver records" the lawyer was no longer interested in the information and I am sure was kicking himself for having asked the question that revealed that I knew what I was talking about.

During the cross-examination, all kinds of questions can be raised. It is very difficult to predict what you will be asked, even by the lawyer for whom you are working. One question which is frequently raised relates to whether you have actually measured the reaction time (or visual acuity or contrast sensitivity, or whatever ability you have commented on) of the person in question. Such a question is intended to discredit your evidence. In most cases you can only respond that you have not measured it, but that based on your knowledge of human performance it is a reasonable assumption that the reaction time of the person in question is within the range of that measured in experiments involving numerous subjects.

Re-Direct

Once you have been cross-examined, your own lawyer is given the chance to clarify any points raised during the cross-examination, called the "redirect." The lawyer can only raise points of clarification and not new information. Then the process is complete.

After Testifying

Once you have given your testimony, you usually depart. The court continues examining any other experts. (Witnesses other than experts usually appear first.) The lawyers make their closing arguments to the judge and jury. In a criminal case, the judge charges the jury, explaining to them the point of law involved and the charge on which they decide guilty or not guilty. In a civil case, a jury usually decides on responsibility for damages. Occasionally, you may be asked to provide questions based on which your lawyer can cross-examine the expert on the opposing side. Usually your part in the case will end after your testimony.

Participating in a court case can be immensely satisfying. In a criminal case, your testimony may help someone whom you feel is innocent avoid a jail term. In a civil case, your testimony can result in enormous financial consequences to one side or the other. Even more important, your testimony can result in relief for the involved parties, providing an explanation of why they made the mistake they did, and why the crash was unavoidable. On your part, participation requires very careful work and typically a real battle of wits with a well-prepared opposing lawyer in public. It is not for the faint of heart.

Appendix: A Day on the Stand [sample notes / observations I took]

Spent from 10:30 a.m. to 4:30 p.m. on the stand. As I waited to go in was wondering why I hated doing this so much. I think it is because the situation is totally out of my control and I know that the opposing lawyer will find errors and holes in my argument, and parade them for all to see. This was not a case for which I felt well prepared – site visit one week ago and report the next day, with no second reader! Did not feel adequately prepared for site visit – insufficient time to work out what I wanted to do and record all relevant observations – in particular the lighting details of the truck used to simulate the accident.

Exam-in-chief was an hour, and Jim, the lawyer, was very good. Used my picture of the dog to get across the idea of needing to have a sense of what was ahead to connect the dots. Cross exam started at 11:45 a.m. after a half hour break. I was accompanied back into the court by a court assistant who whispered to me – "you are such a smart lady" – a reassuring thing to hear on the way to the lion's den. Crown attorney spent a lot of time on the eye exam trying to get me to say my client's vision was poor. Also assumed that poor red/green vision would mean he couldn't tell the difference between headlights and taillights. I'm not sure why she kept on about his visual deficiencies since it certainly didn't help her case. She did that sort of thing several times. Also tried to make a big point of the fact that I was tired when I did the site visit and Andrew was wide awake when the accident occurred – the relevance of this was not clear to me. (I was so thankful I had asked Bruce and Lucy the previous night about Andrew's sleep and usual wake-up time.) The usual criticisms that I didn't test Andrew's PRT.

Broke for lunch at 1:00 p.m. and back again at 2:30 p.m. Tom was there with me for lunch, which was very pleasant – he came to testify with the all-important speed study showing the average was 80 km/h as opposed to the 60 km/h posted. Thankfully they finished with Tom within about 45 minutes.

Parents and the accused were very appreciative of my patience during the erratic and aggressive cross-exam. I feel for them – Lucy told me the approximate cost would be $80,000–$90,000 and there is no recovery. Since our driver did not have a permanent injury, he can't sue.

Jim got up and asked several questions in the re-exam – all about the lights on the truck we tested, which I evaded as best I could since I didn't remember exactly what lights were on the trailer. To my horror he brought up new stuff, to which she objected and as a result got to ask more questions about the lights!!! Finally, at 4:30 p.m. it ended – as with my previous experience of these things the end came rather abruptly.

I must have gone through 20 glasses of water – a very kind court assistant kept sending them my way. When I saw her with a full pitcher, I suddenly realized how much I had gone through.

An exhausting day. Glad it's over.

Index

M

N

O

P

R

W

Y

For Product Safety Concerns and Information please contact our
EU representative GPSR@taylorandfrancis.com Taylor & Francis
Verlag GmbH, Kaufingerstraße 24, 80331 München, Germany